International Conference on Smart Energy and New Power Systems (SENPS 2025)

Journal of Physics: Conference Series Volume 3036

Bangkok, Thailand
9-11 May 2025

Editor:

Shengqing Li

ISBN: 979-8-3313-2268-7
ISSN: 1742-6588

Printed from e-media with permission by:

Curran Associates, Inc.
57 Morehouse Lane
Red Hook, NY 12571

proceedings.com

Some format issues inherent in the e-media version may also appear in this print version.

Institute of Physics
Dirac House, Temple Back
Bristol BS1 6BE UK

Phone: 44 1 17 929 7481
Fax: 44 1 17 920 0979

techtracking@iop.org

Additional copies of this publication are available from:

Curran Associates, Inc.
57 Morehouse Lane
Red Hook, NY 12571 USA
Phone: 845-758-0400
Fax: 845-758-2633
Email: curran@proceedings.com
Web: www.proceedings.com

TABLE OF CONTENTS

Journal of Physics: Conference Series **3036** (2025) 011001 doi:10.1088/1742-6596/3036/1/011001

Editorial

It is a great pleasure and honor for us to welcome you to International Conference on Smart Energy and New Power Systems (SENPS 2025). SENPS 2025 focuses on the field of energy and electrical engineering, which aim to promote the innovative development of smart energy and new power systems at home and abroad. SENPS 2025 exchanges and shares successful innovation experiences in the above fields, and has established an effective community for institutions and industries.

In order to collect more papers, we invited scientists, engineers, educators and students from all over the world to exchange their current technology application experience and research results. This has also promoted the development of relevant fields.

Finally, we have received a good number of responses of around 60 papers from researchers, scholars and industry practitioners covering a wide range of application areas. According to their quality, degree of conformity with our topics and other indicators, we finally accepted 9 papers.

Thank you for your continuous attention and support. We hope this proceeding will provide valuable insights and serve for new scholars interested to explore smart energy and new power systems in various sectors. We believe SENPS conferences will continue to be successful this year and in the future.

Editor:
Prof. Shengqing Li

Journal of Physics: Conference Series **3036** (2025) 011001 doi:10.1088/1742-6596/3036/1/011001

Conference committee

General Chair

Prof. Mohammad S. Obaidat, University of Jordan, Jordan
Dr. Gregory Joseph Duns, AIRCHEM CONSULTING & RESEARCH, Canada

Program Chair

Dr. Elingarami Sauli, Nelson Mandela African Institution of Science and Technology, Tanzania
Dr.Matt Bawn, Institute of Earlham and Institute of Quadram, USA
Dr. Miranda Byrne-Steele, iRepertoire Inc, USA
Dr.Taoyang Wu, University of East Anglia, UK

Publication Chair

Prof. Shengqing Li, Hunan University of Technology, China

Organizing Chair

Dr. Jinhui Zeng, Hunan University of Technology, China
Dr. Zheng Lan, Hunan University of Technology, China
Prof. Fujun Ma, Hunan University, China
Prof. Jianqi Li, Hunan university of Arts and Science, China

International Advisory Chair

Prof. Francisco de León, New York University, USA
Prof. Innocent Kamwa, Laval University, Canada

Program Review Committee

Prof. Vladicescu Popentiu, City University, UK
Prof. Guangzhi Qu, Oakland University, USA
Prof. Dr. Zhengtao Yu,Kunming University of Science and Technology, China
Prof. V.S.S. Yadavalli, University of Pretoria, South Africa
Prof. Bruno Apolloni, Università degli Studi di Milano, Italy
Prof. Harry Bouwman, Delft University of Technology, Netherlands
Prof. Shyi-Ming Chen, National Taiwan University of Science and Technology, Taiwan
Prof. Yahaya Coulibaly, University Technology Malaysia, Malaysia
Prof. Ing Kong, RMIT University, Australia
Prof. Gerald Midgley, Centre for Systems Studies, Business School, University of Hull, UK
Prof. Khubaib Ahmed, Hamdard University, Pakistan
Prof. Moustafa Mohammed Eissa, Faculty of Engineering-Helwan University, Egypt
Dr. Fernando Boronat Seguí, Universitat Politecnica de Valencia, Spain

Journal of Physics: Conference Series **3036** (2025) 011001 doi:10.1088/1742-6596/3036/1/011001

Peer Review Statement

All papers published in this volume have been reviewed through processes administered by the Editors. Reviews were conducted by expert referees to the professional and scientific standards expected of a proceedings journal published by IOP Publishing.

- **Type of peer review:** Single Anonymous
- **Conference submission management system:** Morressier
- **Number of submissions received:** 60
- **Number of submissions sent for review:** 60
- **Number of submissions accepted:** 9
- **Acceptance Rate (Submissions Accepted / Submissions Received × 100):** 15
- **Average number of reviews per paper:** 2
- **Total number of reviewers involved:** 8
- **Contact person for queries:**
  Name: Shengqing Li
  Email: prof_lsq@126.com
  Affiliation: Hunan University of Technology

A LSTM algorithm for accident mechanism of solid oxide fuel cell

Liang Tian*, Fan Yang and Xu Yan

CNOOC Gas & Power Group, Beijing, China

Corresponding author's email: tianliang2@cnooc.com.cn

Abstract. The background of the research on the accident mechanism of solid oxide fuel cell (SOFC) is that with the wide application of SOFC in the field of energy conversion, the safety problem becomes increasingly prominent. It is very important to study the accident mechanism of SOFC to improve its reliability and safety. In today's society, solid oxide fuel cell (SOFC) accidents have the problem of low safety and low accuracy, which greatly affects the safety of solid oxide fuel cells. LSTM algorithm in neural network is an effective accident safety technology for solid oxide fuel cell (SOFC). In this paper, LSTM algorithm is used to create an accident mechanism system, which greatly improves the accuracy and safety of solid oxide fuel cell (SOFC), and reduces the time of management and allocation. Finally, the experimental results show that the LSTM algorithm is easy to operate and the accuracy rate is 95.9%.

1. Introduction

The accident mechanism that SOFC may face during operation mainly includes material failure caused by thermal stress, rupture of electrolyte layer, out-of-control REDOX reaction of electrode material and aging of sealing material. These accidents will not only affect the performance and life of SOFC, but also may cause serious safety accidents such as fire and explosion. Therefore, a thorough understanding of the accident mechanism of SOFC is essential to improve its safety and reliability. Through comprehensive analysis of SOFC material characteristics [1-3], structural design, operating conditions and external environment, the root cause of the accident. For example, by optimizing the material selection and preparation process, the thermal shock resistance and mechanical strength of SOFC can be improved; By improving the structure design, thermal stress concentration can be reduced and electrolyte layer rupture can be avoided. By precisely controlling the operating parameters, excessive oxidation or reduction of the electrode material can be prevented; Through regular inspection and maintenance, aging sealing materials can be discovered and replaced in time. In short, to study the accident mechanism of SOFC deeply is the key to ensure its safe, stable and efficient operation.

Neural networks, especially deep neural networks, build a multi-level, non-linear complex network structure by simulating the connection between human brain neurons. This structure enables neural networks to learn and extract deep features from data, thus achieving remarkable results in image recognition, speech processing, natural language understanding and other fields. However, traditional neural networks often face the problem of disappearing or exploding gradients when processing sequence data, which limits their application in tasks such as time series analysis and machine translation.

This solves the gradient problem of RNN in long sequence training effectively by introducing. The memory unit is responsible for storing long-term state information, while the gating mechanism controls the flow of information, deciding what information should be retained and what should be forgotten. This design allows LSTM to maintain the stability of gradients while working with long series of data, thus capturing dependencies over longer distances.

In addition to theoretical innovation, LSTM algorithm [4-6] also shows great potential in practical applications. In the field of natural language processing, LSTM is widely used in machine translation, sentiment analysis, text generation and other tasks. By learning large amounts of text data, LSTM is able to understand and generate natural and fluent language expressions, greatly improving the accuracy of machine translation and the quality of text generation. In addition, in the field of speech recognition, LSTM also plays an important role. It can accurately capture the timing information in the speech signal. LSTM also performs well in financial fields such as time series forecasting and stock price analysis, providing investors with strong decision support.

It is worth mentioning that with the continuous development of deep learning technology, LSTM algorithms are also constantly evolving and optimizing. For example, in order to solve the memory consumption problem of LSTM when processing very long sequences, researchers have proposed improved methods such as bidirectional LSTM and attention mechanism [7-8]. These improvements not only improve the performance of the model, but also broaden its application range.

In today's era of rapid technological change, artificial intelligence technology is changing our lives at an unprecedented rate. Among them, as one of the core driving forces of image recognition and natural language processing due to its unique structure and strong learning ability. As a special recurrent neural network.

A neural network [9-10] is a computational model that simulates the connections of neurons in the human brain. It consists of a large number of nodes (or neurons) connected to each other. These nodes transfer and process information through weights and activation functions, so as to realize the learning and understanding of the input data. The power of neural networks is their ability to automatically learn features from data without human intervention. This ability makes neural networks highly flexible and adaptable when dealing with complex problems.

However, traditional neural networks have some problems in processing sequence data, such as gradient disappearance and gradient explosion. To solve these problems, LSTM came into being. By introducing gating mechanism, LSTM effectively solves the long-term dependence problem of traditional neural network in processing sequence data. Specifically, the LSTM contains three gates: input gate, forget gate, and output gate. By controlling the flow of information, these gates enable the LSTM to selectively remember or forget certain information, thereby better capturing long-term dependencies in sequence data.

In the process of establishing SOFC system, neural network and LSTM algorithm also play an important role. SOFC system is an efficient energy conversion system, which converts chemical energy directly into electric energy, and has the advantages of high efficiency and environmental protection. However, the design and optimization of SOFC systems involves several factors, such as material selection, structural design, operating conditions, etc., it is necessary to conduct accurate modeling and simulation.

In this process, neural networks and LSTM algorithms can play a key role. Firstly, neural network can be used to model the input-output relationship of LSTM algorithm can be used to model and simulate the dynamic behavior of SOFC system, and analyze the behavior characteristics of the system in different time scales. In addition, the neural network and LSTM algorithm can be used to diagnose and predict the fault of SOFC.

In short, neural network and LSTM algorithm. They can not only improve the design efficiency and accuracy of SOFC system, but also provide strong support for the optimization and improvement of SOFC system [11-13]. it is believed that neural networks and LSTM algorithms will play a greater role in more fields and make greater contributions to the progress and development of human society.

LSTM algorithm is establish the accident mechanism research model of solid oxide fuel cell, and many experiments are carried out, which greatly reduces the probability of solid oxide fuel cell accident and increases the safety. At the same time, the algorithm improves the safety of solid oxide fuel cells.

2. Methodology

SOFC is divided into panel SOFC and tube SOFC. Figure 1 shows PlanerSolid Oxide Fuel Cell (SOFC). The plate SOFC consists of a sandwich structure consisting of an anode, a cathode and a solid oxide electrolyte, in which the anode and the cathode are respectively coated with a catalyst layer, and the electrolyte is usually zirconia (Zr0z) or yttria-stabilized zirconia (YSZ). Tubular SOFC (Tubular Solid Oxide Fuel Cell) group is the mainstream SOFC configuration with short current path, small internal resistance and high volume power density. Tubular Solid Oxide Fuel Cell is shown in Figure 1: it is also composed of an anode, cathode and solid oxide electrolyte, and is not substantially different from the plate type in terms of material. In tubular SOFC, the anode, cathode and electrolyte are arranged separately into a tubular structure, hence the name. The design of the tubular structure can make the reaction gas flow and uniform distribution better, and the aspect ratio of the tubular battery can be relatively large, which increases the surface area of the fuel and the anode contact, improving the fuel utilization rate. The tubular structure reduces the sealing edge, and the effective battery area of the tubular battery with the same sealing perimeter is greatly increased compared with that of the button battery, the current of the single battery is significantly increased, and the reaction gas is more, thus creating favorable conditions for chemical composition analysis such as gas chromatography. Therefore, tubular cells are more suitable for the research of carbon-based fuel cells that need reforming. Of course, the current collection of tubular batteries is more difficult, the current path is long, the internal resistance is large, and the power density is low.

Whether it is a flat or tubular SOFC, the working principle is the same. Taking proton conductor electrolyte batteries as an example, when hydrogen or other fuel flows into the SOFC through the anode, it oxidizes on the anode, producing electrons and protons, and the protons are transmitted through the electrolyte to the cathode, where they chemically react with oxygen on the cathode side to generate water vapor.

Figure 1. Physical diagram of tubular fuel cell.

In this paper, the SOFC-CHP system adopts tubular fuel cell, and the working principle of tubular SOFC is shown in Figure 2.

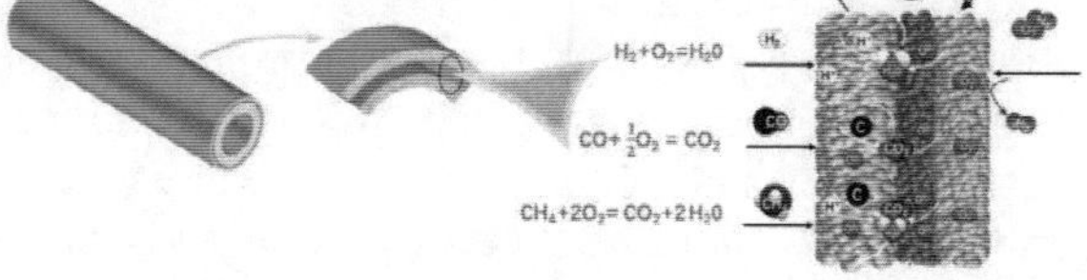

Figure 2. A fuel cell works.

The fuel cell built in Aspen Plus software is shown in Figure 3. Taking pure methane fuel solid oxide fuel cell as an example, the electrochemical module generated in the fuel cell adopts the built-in Fortran language to build, which is more convenient, time-saving and accurate. That is, the unit operation module in the software is directly used to calculate the thermal balance and mass balance of the SOFC stack reaction in the direction of the minimum Gibbs white energy, and the voltage is obtained by including the ohmic loss, activation loss and polarization loss. The SOFC anode can be simulated using a Gibbs reactor (RGibbs), which obfuscates equilibrium and chemical equilibrium calculations using a non-stoichiometric method based on the minimum Gibbs free energy. An electrochemical reaction occurs in the module, and the fuel gas in the inlet stream fully releases heat during the reaction, and the output temperature is controlled to 1000℃. The chemical energy is directly converted into electric energy, and the remaining heat enters the burner with the tail gas.

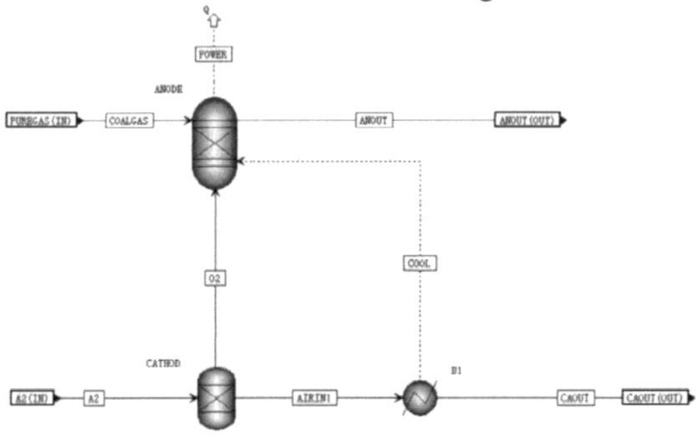

Figure 3. Model construction of SOFC.

A transfer separator (Sep) can be used in the software to simulate the mass transfer process of the battery cathode to transfer oxygen ions through the electrolyte to the anode. The Calculate module in Aspen Plus software writes Fortran language to calculate the 02 amount of fuel required in SOFC, calculate the amount of air required through the air excess ratio, the air passes into the separator, and set the 0 shitter flow rate through the anode in the separator (Sep) to 1. The other groups are divided into zero, simulating the transport of oxygen ions. In this model, the required gas volume is specified according to the fuel utilization rate of the battery, and is transmitted to the anode for electrochemical reaction with the reforming gas (H). The remaining high temperature air is heated by the heat exchanger module in the reformer module, and the remaining heat is discharged with the gas and heated by the heat exchanger.

In this model, the electrochemical module is mainly modeled. Based on the charge conservation equation, its specific construction equation 1 is as follows:

$$I = 2 \cdot F \cdot n_{H_2,EQU} \cdot U_f \tag{1}$$

F is the Faraday constant, whose value is 96487 C/mol, n is the hydrogen equivalent consumed by the SOFC, U is the of the SOFC, that is the ratio of the part of the SOFC that is completely converted into electricity to the input fuel.

In the electrochemical reaction, only H is considered to participate in the reaction in the model, then the hydrogen equivalent is calculated as shown in formula 2:

$$n_{H_2,EQU} = 4 \cdot n_{Fuel} \cdot X_{CH_4} + n_{Fel} \cdot X_{CO} + n_{Fuel} \cdot X_{H_2} \tag{2}$$

Concentration polarization loss: This loss is that with the progress of the reaction, the fuel in the electrode gas diffusion effect is not good, resulting in a low concentration of reaction gas near the reaction interface, and affects the chemical reaction process caused by voltage loss.

At SOFC operating temperature (T_{SOFC}), the reversible voltage (Nerst) is calculated as shown in formula 3:

$$`V_N = V_{N(T_{SOFC})} + \frac{Rg \cdot T_{SOFC}}{ne \cdot F} ln \frac{P_{H_2}\sqrt{P_{O_2}}}{P_{H_2O}} \tag{3}$$

V is Nernst voltage, $V_{N(T_{SOFC})}$ is the electromotive force at SOFC standard condition, and Rg is the gas molar volume constant Rg =8.3145J/molK. n is the number of electron transfer of the chemical substance participating in the reaction in the SOFC, and AG is of the reaction in the battery. ρ_k , α_k , r_K only, is the hydrogen in the simulated system battery.

Partial pressure, partial pressure of oxygen and partial pressure of water vapor.

Voltage polarization loss occurring during battery operation is calculated as follows: Fuel cell ohmic polarization loss V_{0h_T} is calculated as formula 4-6:

$$V_{0\,h\_T} = J \cdot \Sigma_K r_K \tag{4}$$

$$r_K = \frac{\rho_K \cdot \delta_K}{A} \tag{5}$$

$$\rho_k = \alpha_k \cdot exp\left(\frac{b_k}{T_{OFC}}\right) \tag{6}$$

r_k is the ohmic resistance of each part of the battery (Ω), p is the resistance coefficient of the fuel cell component (m·Ω).

Are the average thickness of each battery component (m), α and δ are the ohmic polarization coefficients of the battery, and their values are shown in Table 1.

Table 1. Parameters required for calculating Ohmic voltage loss

Battery module	α_k	b_k	δ
anode	0.0000298	-1392	0.0001
cathode	0.0000811	600	0.0022
electrolyte	0.000029	10350	0.00004
Internal connector	0.000029	4690	0.000085

3. Experiment

The battery system model built in this paper adopts the tubular SOFC stack parameters initiated by Siemens Westinghouse Power Company. In order to verify the accuracy of the system, charge conservation of the SOFC module of the core component is verified in this study. The actual voltage is selected as the indicator, and the battery operating pressure is 1.2bar. The fuel utilization rate is 0.85. The simulation is carried out using the same physical property parameters as in the literature. The direct voltage calculation method is adopted, and the calculated current density is between 1000-5000(A/m?). The output voltage calculated in this paper is very close to the experimental data in the literature, as shown in Figure 4, the maximum maximum mean square error between the output voltage data in the literature and the output data of the simulation calculation in this paper is less than 4.2%, which proves that the SOFC model established in this paper is reasonable.

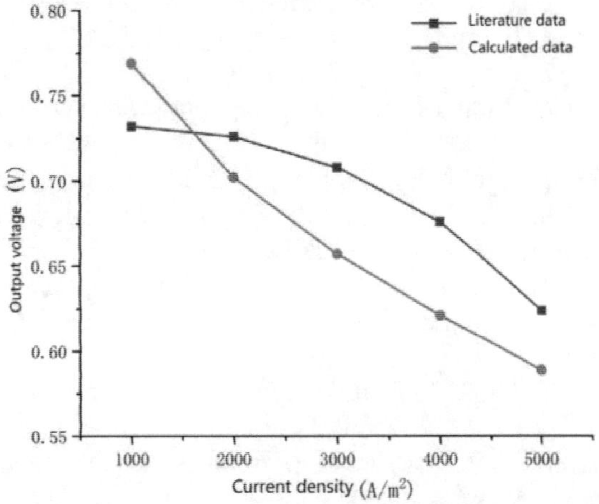

Figure 4. Comparison of the calculated data and the literature data.

In this paper, the accident mechanism of solid oxide fuel cells is compared between the conventional algorithm and the LSTM algorithm, as shown in Table 2.

Table 2. Comparison of accident mechanism algorithms for solid oxide fuel cells

Algorithm	Ideal data	Actual data	Accuracy rate
LSTM algorithm	4392	4216	95.9%
General algorithm	4678	4359	93.2%

4. Discussion

This chapter mainly establishes and verifies the solid oxide fuel cell of the heart component of the SOFC CHP system by using the LSTM algorithm. The calculation of polarization loss is carried out by direct calculation method. The rationality and safety of fuel cell model is verified by comparing the operating parameters of the battery in previous literatures with its simulation results. The SOFC model can be used as a subsystem to integrate other thermal cycling subsystems to model the SOFC CHP system [14].

5. Conclusion

solid oxide fuel cell (SOFC) has been widely concerned because of its high efficiency and environmental protection. However, with the continuous expansion of its application, the security problem of SOFC is becoming more and more prominent. It is very important to study the accident mechanism of SOFC to improve its reliability and safety. In this paper, the LSTM algorithm is used to construct the SOFC accident mechanism system, which significantly improves the accuracy and safety of SOFC, and reduces the time of management and allocation. The experimental results show that the LSTM algorithm is easy to operate and the accuracy reaches 95.9%. In this chapter, the mathematical model of SOFC CHP core components is established and verified by LSTM algorithm, and the polarization loss is calculated by direct calculation method. By comparing the operating parameters of the battery with the data in the literature, the rationality and safety of the SOFC model are verified. The SOFC model can be integrated into other thermal cycling subsystems to simulate SOFC CHP systems.

In the future, with the continuous progress and innovation of technology, this paper believes that there will be more opportunities for cross-integration and mutual promotion in these two fields. This paper can further optimize the performance and efficiency of LSTM algorithm by using machine learning method. At the same time, the idea of LSTM algorithm can be used for reference to solve some complex problems in machine learning.

References

[1] Cheng J, Wang Z and Pollastri G 2021 *Book A neural network approach to ordinal regression (IEEE, 2021, edn.)* pp 1279-1284.
[2] Lumini A and Nanni L 2018 *Current pharmaceutical design* **24** pp 4007-4012.
[3] LeCun Y, Bengio Y and Hinton G 2022 *nature* **521** pp 436-444.
[4] Göller A H 2020 *Drug Discovery Today* **25** pp 1702-1709.
[5] Kumar K V, Ravi V, Carr M and Kiran N R 2008 *Journal of Systems and Software* **81** pp 1853-1867.
[6] Roccetti M, Delnevo G, Casini L and Cappiello G 2022 *Journal of Big Data* **6** pp 1-23
[7] OparC, Wei B and Chen Y 2020 *Book HTMLPhish: Enabling phishing web page detection by applying deep learning techniques on HTML analysis (IEEE, 2020, edn.)* pp 1-8.
[8] Tajik A J 2021 *Book Machine learning for echocardiographic imaging: embarking on another incredible journey (American College of Cardiology Foundation Washington, DC, 2021, edn.)* pp 2296-2298.
[9] Shackelford G and Karplus K 2023 *Proteins: Structure, Function, and Bioinformatics* **69** pp 159-164.
[10] Ortiz-Rodríguez J, Alfaro A R, Haro A R, Viramontes J C and Vega-Carrillo H 2022 *Radiation physics and chemistry* **95** pp 428-431.
[11] Sleeman D 2022 *Communications of the ACM* **29** pp 840-841.
[12] Chakraverty K S and Chakraborty P 2020 *Journal of Educational Technology Systems* **49** pp 170-198.
[13] Mao J, Xu W, Yang Y, Wang J, Huang Z and Yuille A 2014 *arXiv preprint arXiv* pp1412.6632.
[14] Hao X, Zhang G and Ma S 2021 *International Journal of Semantic Computing* **10** pp 417-439.

Journal of Physics: Conference Series **3036** (2025) 012002 doi:10.1088/1742-6596/3036/1/012002

Study on particle size soft measurement of CSTR synthesis process of lithium battery ternary precursor material

Yang Liu[1,2]*, Xueyi Guo[1], Xiaojun Zhuo[2], Hui Tong[1], Xinyi Xiao[2]

[1]School of Metallurgy and Environment, Central South University, Changsha, China
[2]Changsha Mining and Metallurgy Research Institute Co., Ltd., Changsha, China

*Corresponding author's email: liuyang0730@qq.com

Abstract. The precursor particle size distribution has some influence on the calcination process of ternary cathode materials and the performance of the finished product, Its synthesis process was performed using the CSTR (Continuous Stirred-Tank Reactor, Continuous stirring reactor) has the closure of multifeed production and the intermittently of sampling analysis, By analyzing the influence of the multiple production process conditions on the synthetic particle size distribution, Process data collection, input variable selection, standardized processing, and comparison of soft measurement model establishment and comparison, The particle size soft measurement method for forming the precursor synthesis process of lithium battery plant, the results show that the MSE of BiLSTM-GRU particle size D50 prediction by precursor synthesis process was 0.3481, R^2 was as high as 0.9688, and the accuracy was 94.35%, The accuracy rate can reach more than 93%, It has certain production guiding significance.

1. Introduction

Power battery is the 'heart' of new energy vehicles, which is the key to the development of new energy vehicle industry. With the global petrochemical energy supply problems, environmental protection requirements continue to be strict, new energy vehicles as a green low-carbon globalisation of the development of transport, power battery industry has been driven by the rapid development of ternary precursors for power battery cathode material preparation, the demand is growing rapidly. The synthesis of ternary nickel-cobalt-manganese precursors largely determines the physicochemical properties of lithium battery cathode materials, so the research of precursors has received more and more attention. Due to the existence of different particle morphology and uneven particle size distribution during the synthesis of CSTR (Continuous Stirred-Tank Reactor), the subsequent lithiation and electrochemical performance of the synthesized lithium battery are affected. Therefore, in recent years, the physical and chemical properties of precursors have been improved by means of adjusting the feed rate, material concentration, material flow rate, reaction pH and temperature.

In recent years, the research and application of predictive control at home and abroad are becoming more and more extensive, scholars use neural network model predictive control method to identify the controlled object and construct a predictive model to control the output concentration of reactants in CSTR system, and experiments have verified that the proposed neural network model predictive control method shows a good control effect in controlling the output concentration of reactants in the CSTR system. Baranil ingesan I. proposed a continuous stirred reactor temperature control method based on Elman neural network, which can significantly reduce the integral square error of temperature control and improve the accuracy of control [1].

In this paper, the precursor particle size is taken as the research object, based on the data-driven prediction and optimisation of the precursor synthesis process, which provides a reference for the application of precursor production process optimisation in predictive control of reactor parameters.

2. Overview of precursor synthesis technology

2.1. Precursor synthesis process principle

The mainstream method for the preparation of multi-material precursors in industrial production is the co-precipitation method, which is in essence a variety of metal ions in the precipitant complexing agent under the action of the precipitation equilibrium reaction, for the traditional continuous synthesis process, the reaction process, a variety of raw material solutions are added to the reaction kettle at the same time to carry out the reaction, by adjusting the reaction process of the pH, stirring rate, sulphate, as well as ammonia and alkali flow rate, temperature and so on, so that the reaction process of the Particle size distribution in a stable state, from the reactor overflow slurry product indicators to meet the specifications, continuous overflow into the aging kettle, after aging, and then filtration, washing, drying and so on to get the final product, as shown in Figure 1.

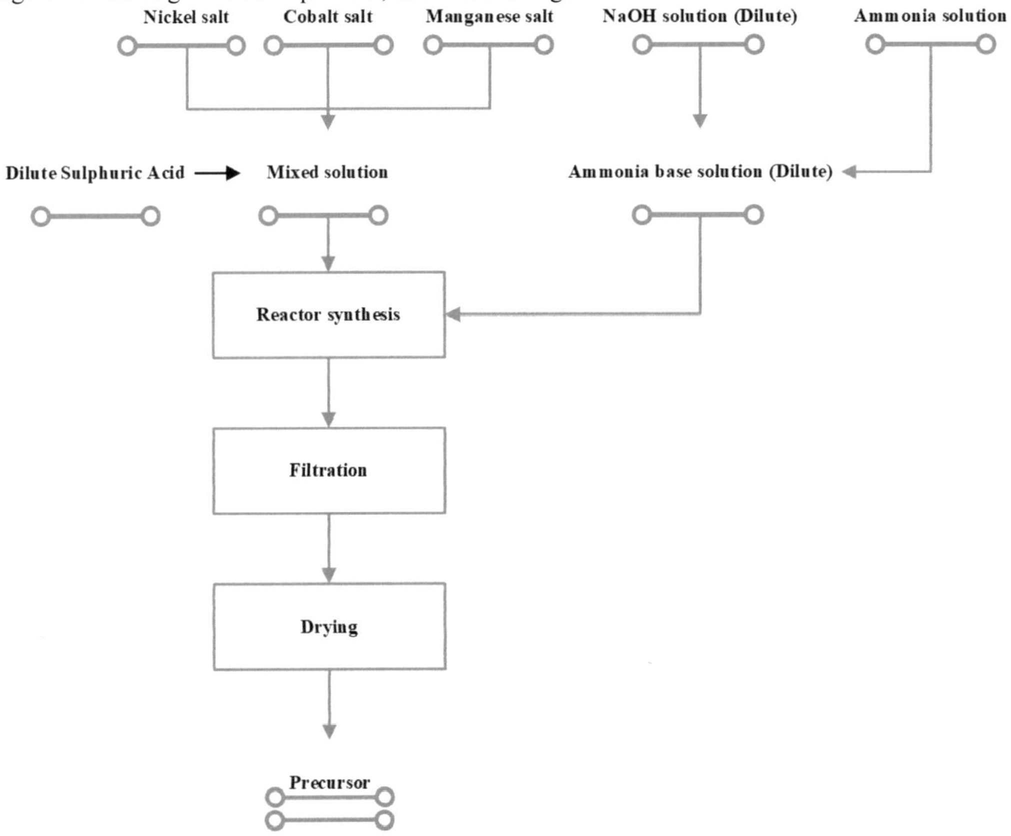

Figure 1. Flow chart of precursor synthesis process

In the above lithium ternary precursor synthesis process, sulfate solution is one of the main raw materials for the synthesis, ammonia as a complexing agent plays a role in the growth of metal particles, while the lye as a precipitant determines the growth of particles faster or slower, pH and temperature is the state of the reaction process feedback, and the value of its parameters is very critical to the growth process of the particles, the shape of the particles is usually need to be frequently and continuously analyzed and observed in order to adjust the feedback of the reaction parameters [2]. Parameter conditions, many companies in the precursor synthesis process, will use a series of high-precision control measures, whether it is the feed solution flow accuracy or the reaction process state quantity, for high nickel 811 or NCA as the type of product, the control accuracy is also more and more demanding.

2.2. Precursor synthesis key factors

Precursor synthesis as a complex complex co-precipitation reaction, its mechanism model has not yet been effectively established, and the particle size distribution as a key production indicator, directly affecting the quality of the product, so it is critical to accurately control the particle size [3]. In recent years, data-driven methods have been widely used in the fields of healthcare, finance, Internet, and electric power by virtue of their ability to transform data into information, integrate and refine the information, and ultimately form automated decision-making models [4]. However, no relevant data-driven research has been seen in the field of lithium ternary precursor synthesis.

During precursor synthesis, process parameters such as sulfate flow rate, ammonia flow rate, lye flow rate, pH, etc. directly affect the particle size distribution, and also determine the morphological growth of particles. The optimisation model based on multilayer feed-forward neural network has powerful nonlinear mapping ability and self-learning ability, which can take each process parameter as input, pass it through the input layer to the implied layer for complex operation, and finally give the prediction result through the output layer [5]. The ability of the basis function in the hidden layer to locally affect the input signal, especially when the signal is close to the centre of the basis function, will produce a larger output, which makes the multilayer feedforward neural network excellent in local approximation computation.

However, conventional multilayer feedforward neural networks have limitations when dealing with time-dependent or dynamically changing data. The process parameters during precursor synthesis vary over time and have time-dependent relationships, and multilayer feed-forward neural networks are usually unable to adequately capture these dynamic changes during prediction [6]. Therefore, recurrent neural networks are more suitable for predicting and controlling complex process parameters of precursor synthesis in practical applications.

3. BiLSTM-GRU prediction model construction

In the process of lithium battery ternary precursor synthesis, particle size prediction is affected by several factors, such as sulfate flow, ammonia flow, lye flow, pH and so on. These factors often show periodic and nonlinear changes, which puts high requirements on the prediction model.

3.1. BiLSTM

The BiLSTM model, with its architecture that combines forward and backward LSTM layers, is able to capture the bidirectional temporal dependence of the data more comprehensively, thus improving the ability to capture complex nonlinear relationships [7]. Compared with the traditional RNN, BiLSTM effectively avoids the problems of gradient explosion and gradient vanishing by introducing state and gating mechanisms, which enhances the robustness of the model in dealing with long time-dependent data. As a result, BiLSTM shows stronger predictive ability and stability in multi-factor granularity prediction of lithium battery precursor synthesis.

Each cell of the LSTM includes four gate operations, the input gate i_t, which indicates the control of whether or not to accept the current input x_t, the forgetting gate f_t, which decides whether or not to retain the state c_{t-1} of the previous time step, the output gate o_t, which decides whether or not to output the current hidden state h_t, and the candidate memory cell $\tilde{C}_t$ is used for updating the candidate values of the state [8]. The computational formulas are respectively:

$$i_t = \sigma(W_i \cdot [h_{t-1}, x_t] + b_i) \quad (1)$$

$$f_t = \sigma\left(W_f \cdot [h_{t-1}, x_t] + b_f\right) \quad (2)$$

$$o_t = \sigma(W_o \cdot [h_{t-1}, x_t] + b_o) \tag{3}$$

$$\tilde{C}_t = tanh(W_C \cdot [h_{t-1}, x_t] + b_C) \tag{4}$$

Where σ is the sigmoid activation function, tanh is the tanh activation function, W_i, W_f, W_o, and W_C are the input weight matrices, and b_i, b_f, b_o, and b_C are the bias terms.

The state is updated using elementwise multiplication:

$$C_t = f_t \odot C_{t-1} + i_t \odot \tilde{C}_t \tag{5}$$

Finally, the computational hidden state is defined:

$$h_t = o_t \odot tanh(C_t) \tag{6}$$

BiLSTM consists of two LSTMs, one processing the input sequence from the forward direction and one from the backward direction [9]. For each hidden state at time step t, the forward LSTM and the backward LSTM compute the hidden state as h_t^{fwd} and h_t^{bwd}, respectively. the final hidden state is a splice of the two:

$$h_t^{bi} = \left[h_t^{fwd}; h_t^{bwd}\right] \tag{7}$$

3.2. GRU

GRU is an RNN variant that reduces the number of parameters and computational complexity of the model by merging the forgetting and input gates in LSTM into a single update gate, while using a reset gate to control how the inputs affect the update of the hidden state. Compared with LSTM, GRU performs more efficiently in dealing with short-term dependencies [10]. Its update formula is as follows:

$$z_t = \sigma(W_z \cdot [h_{t-1}, x_t] + b_z) \tag{8}$$

$$r_t = \sigma(W_r \cdot [h_{t-1}, x_t] + b_r) \tag{9}$$

$$\tilde{h}_t = tanh(W_h \cdot [r_t \odot h_{t-1}, x_t] + b_h) \tag{10}$$

$$h_t = (1 - z_t) \odot h_{t-1} + z_t \odot \tilde{h}_t \tag{11}$$

Where z_t denotes the update gate, r_t denotes the reset gate, $\tilde{h}_t$ denotes the candidate hidden state, and h_t denotes the hidden state at the current time step. W_z, W_r, and W_h are the input weight matrices, and b_z, b_r, and b_h are the bias terms.

3.3. Attention Mechanism

Attention mechanism is a technique used to enhance the effectiveness of the model when processing long sequences by assigning different importance weights to each input element according to the needs of the task at hand. In the process of precursor synthesis, the interaction of multiple factors has a complex temporal relationship on the prediction of granularity [11]. The attention mechanism can help the model automatically focus on the influence of these factors at different points in time, especially when certain factors such as sulfuric acid flow rate and PH value play a key role in particle size change at a specific time period, the model can prioritise these key features. Its main formula is as follows:

$$e_{ij} = a\left(s_{i-1}, h_j\right) \tag{12}$$

$$\alpha_{ij} = \frac{exp(e_{ij})}{\sum_{k=1}^{T} exp(e_{ik})} \tag{13}$$

$$c_i = \sum_{j=1}^{T} \alpha_{ij} h_j \tag{14}$$

Where e_{ij} denotes the unnormalised attention weight, $a(s_{i-1}, h_j)$ denotes the function to compute the attention weight, α_{ij} denotes the normalised attention weight, c_i denotes the context vector, s_{i-1} denotes the hidden state of the previous time step, and h_j denotes the hidden state of the jth time step.

3.4. BiLSTM-GRU Prediction Model Construction Combining Attention Mechanisms

In designing this model, the structure of bidirectional LSTM (BiLSTM) and GRU are combined, the main reason for which is the complementary advantages of the two. BiLSTM is able to better capture long time dependencies and is suitable for dealing with long term variations in the data, whereas the structure of GRU is relatively simple and more computationally efficient, which is particularly good at adapting to short term variations. By combining BiLSTM and GRU, the model achieves a balance between complexity and computational efficiency, while enhancing the flexibility and generalisation of the model. In the task of particle size prediction for ternary precursor particle synthesis in lithium batteries, the data often has multifactorial effects on different time scales. The bidirectional LSTM is responsible for capturing the long-term dependence of these factors, while the GRU efficiently handles short-term feature variations. Combined with the attention mechanism layer, the model is able to dynamically focus on important time steps and key features, which further improves the prediction accuracy. This structural design can make full use of the respective advantages of LSTM and GRU to effectively deal with the complex temporal relationships in lithium battery particle size prediction and improve the overall model performance and computational efficiency. The model structure design is shown in Figure 2.

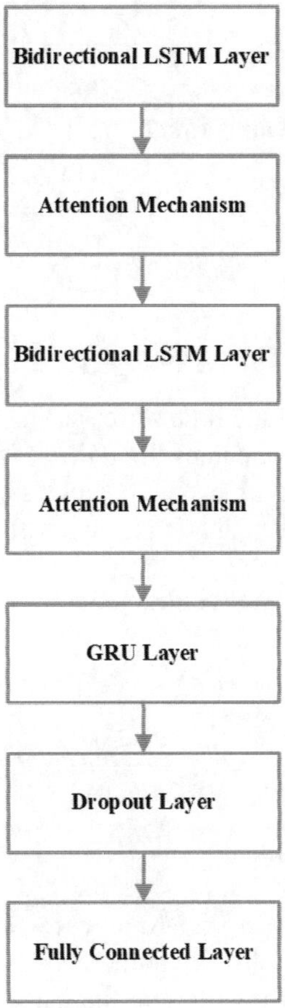

Figure 2. Model structure design

4. Experimental results and analysis

4.1. Data set

In this paper, we use on-site collection and control system production process data and quality inspection and laboratory data, according to the time interval uniform and representative of the 30 sets of data for the training and evaluation of the prediction model, aims to optimize the process control in the process of precursor synthesis through the modelling of the relationship between various types of process parameters and particle size distribution. The data sets included sulfate flow rate, lye flow rate, ammonia flow rate, dopant flow rate, temperature, reactor stirring current value, PH, reactor solid content and particle size. An example of a portion of the data set is shown in Table 1:

Table 1. Partial data set demonstration

Norm	Value 1	Value 2	Value 3
Sulphate flow rate (L/h)	190.04	199.99	600.4
Lye flow rate (L/h)	56.34	83.18	192.22
Ammonia flow rate (L/h)	12.98	18.63	57.91
Doping flow rate (L/h)	100.38	150.43	350.1
Temperature (°C)	71.98	71.95	71.57
Stirring speed (current A)	47.61	42.39	37.94
pH	9.79	9.67	9.84
Reactor solid content	23.09	30.51	156.8
D10	3.68	4.83	7.94
D50	5.23	6.76	11.05
D90	7.386	9.429	15.353

In the data preprocessing stage the data is first normalised with the mathematical expression:

$$X_i = \frac{X - X_{min}}{X_{max} - X_{min}} \tag{15}$$

Subsequently, the moving average method was used to deal with outliers using the following formula:

$$\text{Smoothed Value} = \frac{1}{n}\sum_{i=-k}^{k} x_{t+i} \tag{16}$$

Where, x_{t+i} is the observed value of the time series at time point $t + i$, n is the window size, which is chosen as $n = 3$ in this experiment; i denotes the index of the sliding window from $-k$ to k, which is used to compute the average of the time point t and its previous and subsequent k time steps; k denotes the size of the sliding window, which is chosen as $n = 3$ in this experiment, i.e., the window size is 3.

Finally, the data are reshaped to fit the requirements of the time series model. The data reshaping steps include:

Define the time step: set $T = 1$;

Number of features: the number of features is F;

Number of samples: the number of samples $S = N - T + 1$, where N is the number of samples of the original data.

4.2. Parameter setting and evaluation index

In this study, the dataset is divided into training set and test set according to 7:3 ratio to evaluate the model performance. A 3-fold cross-validation is used to optimise the model hyperparameters and improve its prediction performance. The hyperparameters of the BiLSTM-GRU model were optimised using the grid search provided by Scikit-learn, and the main hyperparameters optimised included the learning rate (lr), the number of LSTM units (lstm_units), the number of GRU units (gru_units), and the dropout rate (dr). The optimal combination of hyperparameters determined by grid search is lr=0.001, lstm_units=8, gru_units=16, and dr=0.1. Finally, predictions are made using this BiLSTM-GRU model and its performance is evaluated on a test set.

The performance of this prediction model is measured by calculating the average mean square error (MSE), root mean square error (RMSE), coefficient of determination (R^2), and percentage accuracy (Accuracy) of the predicted and actual values. The formula for calculating each indicator is as follows:

$$MSE = \frac{1}{n}\sum_{i=1}^{n}(y_i - \hat{y}_i)^2 \tag{17}$$

$$RMSE = \sqrt{MSE} \tag{18}$$

$$R^2 = 1 - \frac{\sum_{i=1}^{n}(y_i-\hat{y}_i)^2}{\sum_{i=1}^{n}(y_i-\bar{y})^2} \tag{19}$$

$$Accuracy = \left(1 - \frac{|y_i-\hat{y}_i|}{y_i}\right) \times 100\% \tag{20}$$

4.3. Model prediction results

The BiLSTM-GRU was used to predict the particles in different time steps, and the prediction results are shown in Figure 3. As can be seen from Figure 3, the model captures the particle size trend accurately, but the change is more drastic in some time steps, which may be due to the more complicated synthesis process of ternary precursors in that time step. For the prediction of smaller particle sizes, the deviation between the predicted and actual values is sometimes larger, which suggests that the present model may need higher resolving power or further feature engineering optimisation when dealing with smaller scale particle sizes.

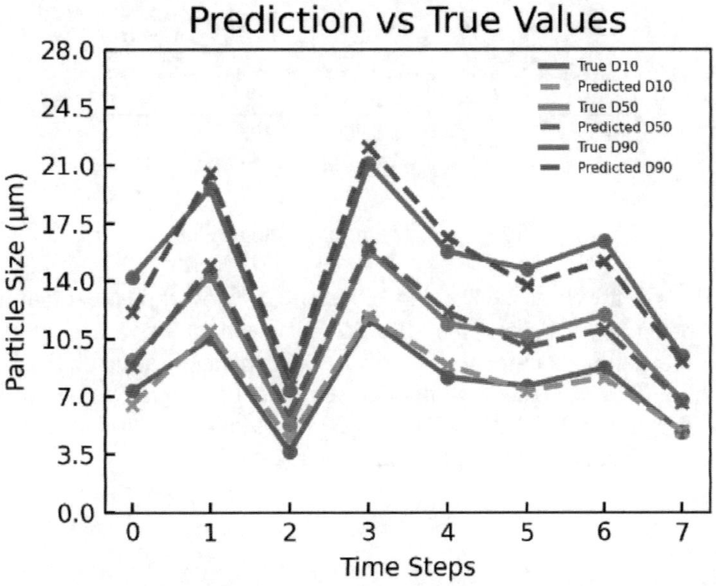

Figure 3. Model prediction effect

Table 2 demonstrates the performance metrics of the model at different prediction granularities, revealing the differences in model performance on different target variables. Specifically, in the D50 prediction task, the model demonstrated the best performance with an MSE of 0.3481, an R^2 as high as 0.9688, and an accuracy of 94.35%. For D10 prediction, despite having the lowest MSE, the R^2 was lower and slightly less effective compared to D50. In conclusion, the model predicts best at D50, followed by D10, and relatively weakly at D90, which may be related to the complexity of the different target variables and the variability of the data.

Table 2. Performance prediction of different target variables

Granularity	MSE	RMSE	R^2	Accuracy(%)
D10	0.3055	0.5527	0.9505	92.86
D50	0.3481	0.5900	0.9688	94.35
D90	1.2830	1.1327	0.9315	92.74

As shown in Figure 4, with the training, the MSE of the training set fluctuates greatly in the early stage, but with the increase in the number of iterations, the training error gradually decreases and tends to be stable, indicating that the model gradually converges. The MSE of the test set and validation set changes steadily and stays at a low level, indicating that the model is stable in its prediction ability and has good generalisation ability.

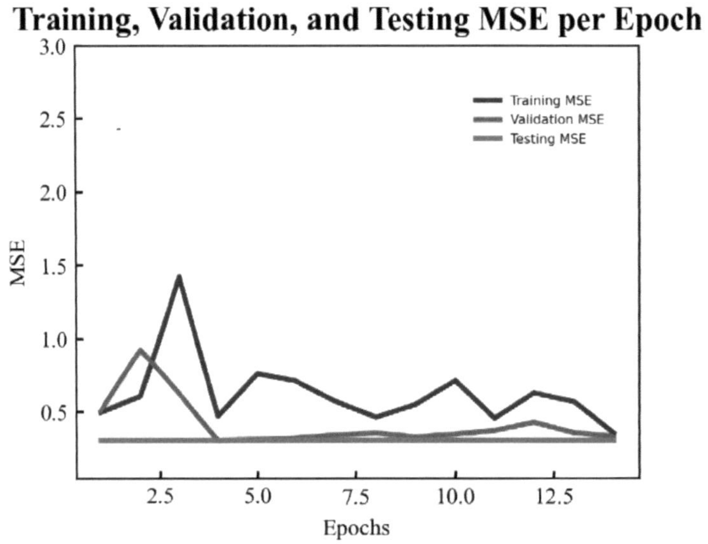

Figure 4. Training of the model

As shown in Figure 5, the R^2 of the training set reaches 0.9867, indicating that it learnt most of the granularity change patterns. Meanwhile the difference between the R^2 values of the validation set and the test set is small, indicating that the model shows a better generalisation ability in the validation and test phases. This is verified in Figure 5. Although the model has some deviations in individual time steps, the overall prediction results for different granularities show its strong adaptability and prediction ability in multi-scale complex granularity change tasks.

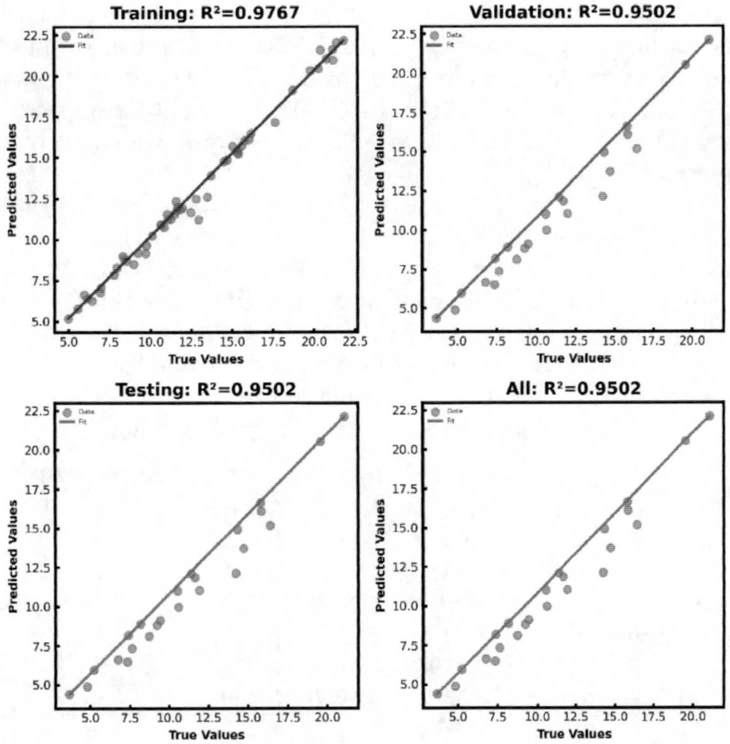

Figure 5. R^2 Comparison

4.4. Evaluation of model effect

In order to evaluate the fitting effect and prediction accuracy of the BiLSTM-GRU model, further comparison experiments with Random Forest, SVR, LSTM, and Transformer models are performed. The figure demonstrates the comparison of the prediction effect of the five models in D50 granularity. As can be seen from Figure 6, the SVR model has relatively poor prediction effect and the BiLSTM-GRU model has better prediction effect.

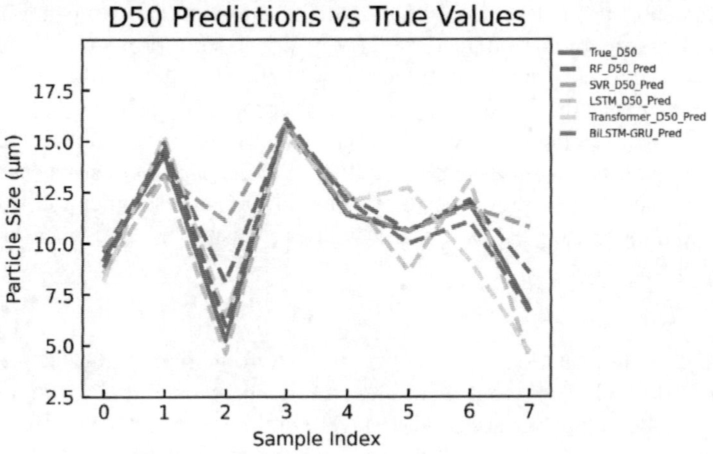

Figure 6. Comparison of D50 prediction results

Table 3 demonstrates the performance of different models in the prediction task. The comparison reveals that the BiLSTM-GRU model outperforms the other models in all indicators. Firstly, BiLSTM-

GRU has the lowest MSE of 0.3481, indicating that the model has the least error in prediction. The coefficient of determination (R^2) is as high as 0.9688, indicating that the model fits the data extremely well. In comparison, the traditional Random Forest model, although performing moderately well, has an MSE of 1.8078, which is not as good a fit as BiLSTM-GRU, while the Support Vector Regression (SVR) model, especially SVR 2, has an MSE as high as 10.8188, which is poorly fitted and struggles to capture data patterns effectively.

Although the Transformer model performs well in processing natural language tasks, its MSE (3.2279) and RMSE (1.7966) are both higher on the small-sample dataset, and its R^2 is only 0.7303, which is lower than that of the LSTM model (0.8037). This suggests that the predictive power of the Transformer is relatively limited in small-sample tasks. In contrast, the BiLSTM-GRU model, thanks to its bidirectional structure combined with GRU, is able to better capture the complex dependencies in the time series, which significantly improves the prediction performance of the model. Therefore, BiLSTM-GRU is undoubtedly the optimal choice for this kind of time series prediction task.

Table 3. Comparison of model prediction results

Model	MSE	RMSE	R^2	Accuracy(%)
Random Forest	1.8078	1.3446	0.8515	87.46
SVR 0	2.9392	1.7144	0.5237	76.02
SVR 1	6.4417	2.5380	0.4219	76.72
SVR 2	10.8188	3.2892	0.4223	77.75
Transformer	3.2279	1.7966	0.7303	83.50
LSTM	2.9796	1.5558	0.8037	85.95
BiLSTM-GRU	0.3481	0.5900	0.9688	94.35

5. Discussion

This study proposes a soft measurement method based on a combination of BiLSTM-GRU and an attention mechanism for predicting the particle size distribution during the synthesis of ternary precursor materials for lithium batteries. By modelling the time series of multiple process parameters such as flow rate, pH value and temperature, the model can effectively capture complex nonlinear and time-series relationships and achieve high prediction accuracy. In the D50 particle size prediction task, the MSE of the model was 0.3481, the coefficient of determination (R^2) was as high as 0.9688, and the prediction accuracy reached 94.35%, which is significantly better than traditional models such as random forests, SVR and Transformer. This shows that the BiLSTM-GRU model takes into account both long-term and short-term dependencies, while the attention mechanism further enhances the model's ability to focus on important features at key time steps, providing more stable prediction results for complex synthesis processes. Accurate particle size distribution prediction is of great significance for improving the quality of precursor products, reducing production costs, and optimising industrial control parameters. However, due to the small sample dataset and feature dimension, the model still has some errors in predicting smaller particle sizes (such as D10). Future research can further improve prediction performance and industrial application results by increasing the data scale, introducing more process features such as pressure and residence time, and optimising the interpretability of the model.

6. Conclusion

The aim of this paper is to provide a particle size prediction method for CSTR production of ternary precursor particles, through the input variables of sulphate, lye and ammonia flow rate, pH, temperature, etc., and the output variables of D10, D50 and D90 particle sizes, for the construction of the prediction algorithm model, and to study the fusion method of BiLSTM-GRU and the attention mechanism, so as to make the prediction results of good convergence, and to compare with other conventional algorithms, with better model evaluation results. By accurately predicting the particle size, it meets the requirements of industrial production control, helps the particle size control to be more stable, and achieves the key role of improving the quality of production products and reducing the production cost.

References

[1] Baranilingesan, I. (2021). Optimization algorithm-based Elman neural network controller for continuous stirred tank reactor process model. *Current Science (00113891)*, *120*(8).

[2] Vaughey, J., Trask, S., & Poeppelmeier, K. (2021). Synthesis as a design variable for oxide materials. *The Electrochemical Society Interface*, *30*(1), 53.

[3] Ling, J., Karuppiah, C., Krishnan, S. G., Reddy, M. V., Misnon, I. I., Ab Rahim, M. H., ... & Jose, R. (2021). Phosphate polyanion materials as high-voltage lithium-ion battery cathode: a review. *Energy & Fuels*, *35*(13), 10428-10450.

[4] Guo, Q., Huang, J., Liang, Z., Potapenko, H., Zhou, M., Tang, X., & Zhong, S. (2021). The use of a single-crystal nickel-rich layered NCM cathode for excellent cycle performance of lithium-ion batteries. *New Journal of Chemistry*, *45*(7), 3652-3659.

[5] Shim, J., Kim, Y. J., & Bang, J. H. (2024). Surface Pinning of Mn by Oxidation State Control for the Synthesis of Cobalt - Free, Ni - Rich, Core/Shell Structured Cathode Materials. *Small*, 2400518.

[6] Xu, J., Gao, J., Wang, M., Tian, S., Lin, C., & **a, Y. (2024). Synthesis of a Ce–La Adsorbent with Large Adsorption Capacity for Efficient Fluoride Removal from Leaching Solution of Spent Lithium-Ion Batteries. *Energy & Fuels*, *38*(8), 7379-7388.

[7] Wang, D., Deng, Y. P., Liu, Y., Jiang, Y., Zhong, B., Wu, Z., ... & Chen, Z. (2023). Sodium-ion batteries towards practical application through gradient Mn-based layer-tunnel cathode. *Nano Energy*, *110*, 108340.

[8] Ben-Barak, I., & Obrovac, M. N. (2024). All-Dry Synthesis of NMC from [Ni, Mn, Co] 3O4 Spinel Precursors. *Journal of The Electrochemical Society*, *171*(4), 040535.

[9] Ou, X., Liu, T., Zhong, W., Fan, X., Guo, X., Huang, X., ... & Lu, J. (2022). Enabling high energy lithium metal batteries via single-crystal Ni-rich cathode material co-do** strategy. *Nature communications*, *13*(1), 2319.

[10] Ashworth, D. J., Driver, J., Sasitharan, K., Prasad, R. R., Nicks, J., Smith, B. J., ... & Foster, J. A. (2023). Scalable and sustainable manufacturing of ultrathin metal–organic framework nanosheets (MONs) for solar cell applications. *Chemical Engineering Journal*, *477*, 146871.

[11] Gómez-Palos, I., Vazquez-Pufleau, M., Schäufele, R. S., Mikhalchan, A., Pendashteh, A., Ridruejo, Á., & Vilatela, J. J. (2023). Gas-to-nanotextile: high-performance materials from floating 1D nanoparticles. *Nanoscale*, *15*(13), 6052-6074.

Data governance method for unmanned aerial vehicle inspection for power grid equipment management

Sijia Zheng*, Xuan Yang, Zecheng Li, Yawen Gong

State Grid Electric Power Space Technology Company Limited, Beijing ,102209, China

*453136965@qq.com

Abstract. Power inspection is gradually shifting towards digitization and intelligence, which puts higher requirements on the quality of basic data. In the power production process, the generated data primarily originates from equipment status and maintenance activities. The suboptimal quality of data from these aspects leads to inadequate performance in analytical applications. To address this issue, this paper proposes a UAV inspection data governance method for power grid equipment management. The method involves acquiring UAV inspection data, including basic records and line video data of ultra-high voltage transmission lines, extracting attribute features of equipment from the videos, and comparing them with the equipment records. A data governance approach based on feature engineering is designed, incorporating data verification algorithms for equipment attribute completeness, topological connectivity, nameplate compliance, and measurement consistency. This enables the validation of maintenance data by integrating real-time measurement data. For missing or erroneous information already present in the system, data governance identifies the root causes, supplements missing data, and corrects errors, thereby resolving inconsistencies across information systems. This method provides a solution framework for enhancing the quality of power grid equipment management data and supporting subsequent analytical applications.

1. Introduction

With the gradual advancement of power grid informational construction and the continuous deepening of production management information system construction, the ledger data, operation data, defect, test, maintenance and other operation and maintenance data of power transmission and transformation equipment are becoming increasingly informational, and the amount of data generated is increasing greatly [1]. These data are closely related to production safety and equipment management, and the standardization, correctness, timeliness, and completeness of data are becoming increasingly important for supporting and making decisions in power grid work. However, due to the fact that production data comes from grassroots power supply units and belongs to second-hand data, many attributes of historical data are not complete, especially the lack or error of key data such as manufacturers, equipment models, and operation time for equipment life-cycle management, which seriously affects the evaluation and analysis of equipment family defects [2,3]. Data quality is determined by a series of parameters such as data consistency, correctness, and relevance. In order to improve the value of data utilization in the production management information system, this paper focuses on the data quality of the production management information system of Inner Mongolia Power Company [4]. By examining equipment ledgers, technical parameters, test reports, maintenance records, defect records, and other information, the equipment data is analyzed from multiple dimensions such as completeness, correctness, and consistency, and solutions are proposed to improve data quality and achieve the reliability of source data throughout the entire life cycle of equipment.

The absence or inaccuracy of equipment technical parameters, especially the lack of critical data, can significantly impact precise equipment evaluation and decision-making regarding operational safety. For instance, incorrect or missing short-circuit withstand parameters of a transformer directly affect its accurate assessment and proper analysis of its actual condition. Missing manufacturer and model specifications hinder data sharing, which is essential for conducting family defect analysis—following the grid equipment family defect management approach — on identical models from different manufacturers. If operational rules, maintenance recommendations, and other general information cannot be shared, it becomes impossible to conduct comprehensive early-warning studies for in-grid equipment. Missing manufacturing and commissioning dates disrupt full lifecycle management, preventing correct analysis of equipment service years, thereby affecting operational and maintenance planning as well as aging equipment monitoring. The absence of pollution level classifications reduces the timeliness of management decisions, such as failing to promptly obtain technical support (e.g., operational and maintenance data) for similar equipment when analyzing faults at a substation. To address missing equipment parameters in UAV inspection data, this study proposes an equipment-identifier-based core node approach to analyze association errors, leveraging equipment correlation results to identify parameter inconsistencies. Compared to traditional data governance methods relying on basic attribute constraint rules, the proposed method considers inter-parameter relationships which improves the accuracy of error detection.

2. Related Work

The validation of equipment parameters involves extensive checks. Focusing on the key requirements of condition-based maintenance (CBM), correctness verification is performed for critical data in CBM tasks, with corresponding inspection rules established. In accordance with the Power Company Production Management Information System Guidelines for Transmission and Distribution Equipment Record Registration, correctness analysis is conducted for over 3 million critical parameters from 210,000 primary substation equipment [5]. The data screening approach prioritizes conventional foundational data for validation. Key attributes such as voltage level, equipment type, and associated bay are primarily selected for equipment condition assessment. For important parameters in the state maintenance process, the inspection rules are based on the expression of voltage level, the relationship between factory date and commissioning date, equipment name and equipment type mismatch, interval attribution, and equipment model description. According to the analysis results, 63.8% of them have the correct production date and commissioning date, 67.2% have the same equipment name and type, and 69.7% have the correct interval [6]. The data in the power equipment production management information system shows a power exponential trend of growth. In practical applications, due to the lack of effective means to check the original data, a large amount of garbage data has been generated. If a large number of errors and duplicate garbage data are found in various records during data integrity checks during the production process, it will cause the system to run slower and slower [7]. During the data reading process, these garbage data are not processed, which greatly confuses the results of data analysis and affects decision analysis.

The governance of electric power maintenance data provides a data base support for achieving unified and intensive management of data, ensuring safe sharing and analysis of data, and also increases the value of data resources. State Grid Corporation of China collects and constructs business metadata from several heterogeneous video image data sources, including data standards, data models, and data quality rules [8]. After completing the classification of heterogeneous data sources, distinguish the basic attributes of video images, video image description information, and video image raw data. For the basic attributes and description information of video images, key data items of these two types of data are verified and analyzed based on the constructed business metadata [9]. For important data with issues discovered during verification, data cleaning methods such as data coverage, data filtering, and data completion are used to ensure the standardization, validity, and correlation of the data. For the basic attributes and description information of video images, data conversion is implemented based on the constructed business metadata, including data standard conversion, data code translation, and data field normalization [10]. For the basic attribute data of video images, the description information data of video images, and the raw data of video images, key attribute governance is carried out in terms of image quality, device time, device latitude and longitude, and abnormal data volume [11]. Video image data has the characteristics of strong temporal and spatial correlation. When time or spatial data is abnormal, it will seriously affect the effectiveness of video image related business applications. In equipment fault identification applications, quality analysis of image data is carried out on collected images or photos to avoid the intelligent image parsing system from processing invalid images or photos. Manage raw data through data governance tools, evaluate quality issues identified during the data governance process, generate evaluation reports, and assist in locating and preventing quality issues.

3. Classification and Integration of Multi-Source Monitoring Data for Transmission Channels

The data monitoring devices in transmission channels include sensors, video monitors, and state sensing devices. There are significant differences in the data collection methods and recording formats among different types of collection devices. Using image identification algorithms to identify device types and other features in video data, comparing the identification results with device types in device archive information, and using the comparison results as the basis for data fusion judgment.

3.1. Collection and Classification of Monitoring Data for Transmission Channels

The monitoring data of transmission channels includes micro meteorological data, monitoring equipment, multi-sensor data fusion, multi-sensor data fusion target recognition principles, infrared thermal imaging, and ultraviolet imaging data. Micro meteorological data is mainly collected by micro meteorological monitoring systems, which extract micro meteorological environmental data of transmission line channels. The system transmits meteorological parameters such as temperature, humidity, wind speed, wind direction, and rainfall from all monitoring locations equipped with monitoring devices to the data center. The data center predicts the climate change characteristics of the power line channel environment based on the fluctuation curve of the above parameters, and makes decisions on lightning arrested installation, pole climbing methods in polluted areas, etc. to reduce the probability of line failures. The video monitoring system monitors the surrounding conditions of transmission lines through video acquisition devices, making it convenient for transmission line monitoring personnel to visually observe changes in the surrounding conditions of the lines. The remote monitoring system installed on the transmission line adopts advanced digital video compression technology, which transmits the surrounding situation of the line to the background monitoring center in real-time through wireless communication, and can set up programs to alarm behaviors that endanger the safety of the line. Adopting infrared detection technology to conduct all-weather monitoring of towers in high-risk areas of transmission lines, in order to timely eliminate hidden dangers of accidents. The application of data classification results can effectively reduce power accidents caused by external forces such as construction around power lines.

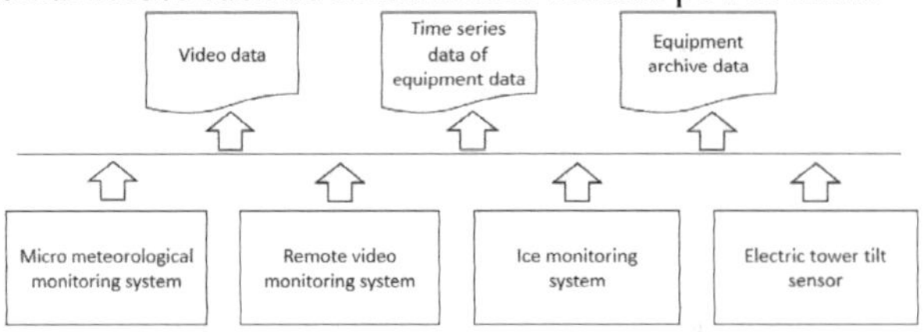

Figure 1. Data collection and classification of transmission channels

As is shown in figure 1, the icing monitoring system installs an automatic monitoring station for icing on iron towers in areas prone to icing. By adopting accurate monitoring and analysis methods and practical mathematical models, real-time monitoring of the icing status of transmission lines can be carried out. It can monitor the icing situation of high-voltage transmission lines and substation insulators operating in harsh atmospheric environments online, timely detect changes in equivalent icing thickness within a vertical spacing unit, and provide users with warning values according to line design standards. The tower tilt sensor will collect data such as lateral tilt, longitudinal tilt, and composite tilt of the tower and send it to the monitoring center through wireless communication or optical cable transmission devices. The monitoring center will store, display, and statistically report the status parameters of lateral tilt and longitudinal tilt, and analyze them in conjunction with the tower's own design parameters to complete the multi parameter warning function of tower tilt.

3.2. Integration and Fusion of Multi-Source Data In Transmission Channels

The multi-source data of transmission channels includes unstructured video and voice data, as well as structured time series data and basic archive data. Data with different structural types cannot be directly fused into a specific type of data, so feature extraction is required for video and sequence data. Identify common descriptive media in video data and sequence data, and map multi-source data to the same data dimension.

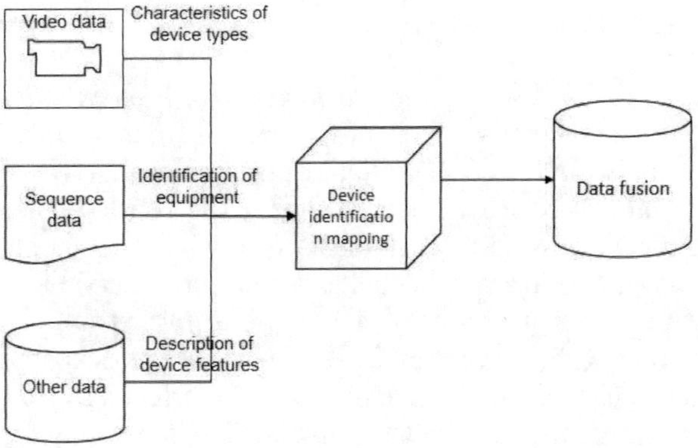

Figure 2. Integration and fusion method of multi-source data in transmission channels

As shown in Figure 2, based on the analysis of multi-source heterogeneous data, define the entity types, attribute types, and entity relationships of transmission channel monitoring data objects. After determining entity types, attribute types, and entity relationships, construct an organizational framework for knowledge of distribution networks. The knowledge graph of multi-source heterogeneous data in transmission channels can be represented by G=(E, R, S), where E is the set of entities in the knowledge base, R is the set of relationships, and S is the set of triplet knowledge. Data layer construction refers to extracting the required entities and relationships from multi-source heterogeneous data in the distribution network under the guidance of the pattern layer organizational framework. Obtain the original transmission channel access data, and use multi-source heterogeneous data fusion technology to summarize, correlate, and integrate various data from different data sources in the distribution network. The fused multi-source heterogeneous data, which includes cleaning, denouncing, and standardizing structured and semi-structured data, to ensure data quality and consistency. Structured transformation, Chinese word segmentation, text annotation, and construction of word vector models for unstructured data. Extract the required entity and relationship information from processed data using rules and deep learning based knowledge extraction methods. Integrate knowledge obtained from different data sources and knowledge extraction methods, eliminate duplication and conflicts, and form a consistent knowledge representation. By utilizing the knowledge storage technology of association models, the extracted and fused knowledge is stored to form a multi-source heterogeneous data knowledge graph for power transmission channels. Use a graph database to

Journal of Physics: Conference Series **3036** (2025) 012003 doi:10.1088/1742-6596/3036/1/012003

store monitoring data of transmission channels, forming a complete and structured knowledge graph. This graph contains information on various entities, attributes, and relationships in the field of distribution networks. Utilize the multi-source data fused in the knowledge graph of the distribution network to monitor the operational status of various equipment and lines involved in the transmission channel in real-time. Quickly locate abnormal measurement data through interactive display of knowledge graphs. Perform data preprocessing and knowledge extraction during the data layer construction process to ensure the quality and consistency of the data. The knowledge graph of multi-source heterogeneous data in distribution networks can integrate data from different systems and analyze the status data of high voltage and medium voltage power lines.

4. Governance and Application of Monitoring Data for Transmission Channels

The quality of transmission channel monitoring data plays an important role in promoting the safety analysis and equipment maintenance analysis of transmission channels. Feature engineering utilizes feature values or vectors to reconstruct or enhance the characteristic representation of data, which has been applied in data governance algorithms to improve the detection probability of erroneous data features. Using feature engineering based data governance algorithms to address the issues of insufficient integrity, consistency errors, and lack of standardization in transmission channel monitoring data. Based on the results of data governance, carry out applications such as transmission channel security analysis and anomaly detection.

4.1. Research on Governance Methods for Monitoring Data of Transmission Channels

The quality issues of videos in transmission channel monitoring data mainly lie in issues such as video clarity, video shooting angle, and video correlation misalignment. The resolution of video clarity issues can be achieved through edge enhancement methods. The edge detection unit calculates the direction of the dominant edge passing through the current pixel in the brightness domain, and then an edge enhanced brightness calculation unit calculates a new current pixel brightness value. Replace the original pixel matrix with the calculated new data values to further improve the edge strength along the dominant edge direction at the current pixel. The edge direction is approximately determined by the direction of brightness change, and the magnitude of the brightness change rate crossing a certain direction is used to approximate the edge intensity in that direction.

The solution to the problem of angle deviation in video shooting can be achieved through deep learning to optimize the target object in the image that has been offset. Using deep learning for object detection and image cropping to obtain the foreground area that needs to be protected, while the remaining areas are treated as background areas. Obtain a distortion correction result map based on distortion correction for the input image, as a reference for the results of the background area mentioned above. Perform a spherical transformation on the distortion correction result map to obtain a spherical transformation result map, which serves as a reference for the results in the aforementioned foreground area. The foreground and background in the image need to have a smooth transition to avoid abrupt changes at the edges of the resulting image. Combined with other constraints, the above problem is transformed into an optimization problem with a transformed grid for solution. After obtaining the transformed grid, interpolation is performed to obtain the final result graph.

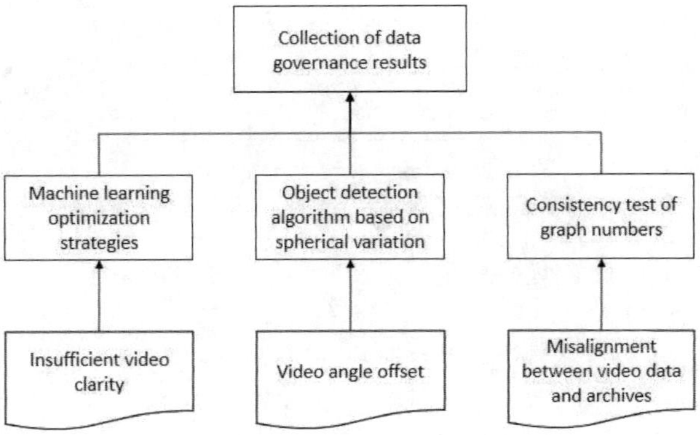

Figure 3. Governance of monitoring data for transmission channels

As shown in Figure 3, in the actual process of image target correction, target detection operations can be combined to determine the presence of target objects during photography. Then determine whether further processing is needed based on its distance from the center of the image and the size of the portrait. The farther away from the center of the image, the more severe the deformation of the target image with a larger area. If not required, only perform distortion correction on the input image. If necessary, perform the aforementioned spherical transformation again after distortion correction. The preview process is similar, but a transition process can be added between the two deformation operations to avoid image shake caused by sudden switching. The core of this method lies in spherical transformation, which can be achieved using existing open source engineering. One characteristic of it is that the output image shrinks non proportionally compared to the input image, and the central area image is magnified to a certain extent. Applying this operation to the distortion corrected image resulted in a certain degree of restoration of the proportion of the figures at the edges in the image due to the shrinkage operation.

The solution to the problem of video association misalignment can be achieved through feature identification methods. Select video data with clear identification of the target object as the sample based on the original data. Then, manual annotation method is used to identify key attributes such as type, number, voltage level, etc. of the target object in the sample set. After the identification is completed, input other video data and use machine learning algorithms to classify and recognize the type, number, voltage level, and other information of the target object in the original video data. After obtaining the recognition results, it will continue to be manually verified and the video data with incorrect annotations will be corrected.

4.2. Application Analysis of Transmission Channel Data Governance

After completing the governance of monitoring data for the transmission channel, obtain the data-set required for safety analysis of the transmission channel. Identify and label each data in the data-set to obtain a labeled training data-set. Extract the features corresponding to any training data in the labeled training data-set, and construct neural network convolution layers and networks based on these features. Reduce the dimensional and residual connections of any feature corresponding to the training data, and combine it with the neural convolution network to construct a learning model. Train the initial security monitoring network alliterative based on the labeled training data-set. And optimize the feature learning module and feature classification module of the neural network in iterative training to obtain the optimized lightweight security monitoring network. Deploy the optimized lightweight security monitoring network to edge devices and conduct real-time security monitoring of transmission channels. The prepossessing and truth labeling of each data in the data-set are carried out, the data-set is filtered and the data containing hidden targets is retained, and then the low-quality data is removed. The poor quality data includes data with blurriness exceeding the threshold, watermarked data, and incomplete target data. Adjust the filtered data to the set size. Label each re-sized training data with truth labels to obtain a labeled training data-set. Deploy the trained security analysis model to the edge devices of the transmission channel monitoring device and conduct real-time security monitoring of the transmission channel.

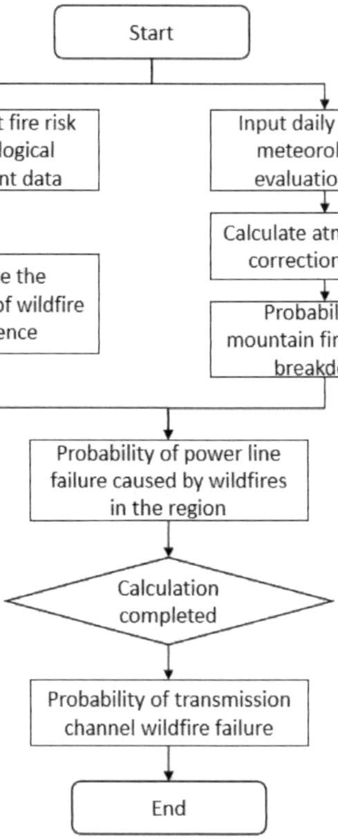

Figure 4. Method for predicting wildfire faults based on data governance results.

Evaluate the meteorological level of the fire through meteorological parameters, and calculate the daily average weather parameters of the area, such as maximum temperature, minimum relative humidity, precipitation, wind speed, etc., through quantitative scoring. After determining the highest temperature index, lowest relative humidity index, precipitation index, and wind power level index, the fire weather index is calculated. Mountain fires not only affect the ambient air, but also produce smoke, ash, and various small particles during the combustion process, which greatly reduces the insulation strength of the gaps, and the breakdown voltage can be reduced to up to 10% of the original voltage. Therefore, the smoke correction coefficient is introduced to obtain the power frequency breakdown probability for calculating the air gap during mountain fires based on the correction of atmospheric correction coefficient and dense smoke correction coefficient.

5. Conclusion

This paper proposes a multi-source data governance method for power grid equipment management to address the problem of large scale but low data quality in transmission channel monitoring data. Using object detection algorithms to annotate the video monitoring data collected by transmission channel devices, and annotating key information such as equipment type and voltage level in the video data object. Using annotation results as samples, construct a neural network feature classifier to classify the video data collected by the transmission channel detection device according to their features. Delete video data with insufficient clarity or abnormal shooting based on classification results to improve the quality of transmission channel monitoring data. This method provides data support for the application of transmission channel safety detection and analysis.

References

[1] Tao X, Xia Y, and Chen J. Research on multispectral remote sensing image enhancement method of intelligent patrol UAV 2025 *J. Northeast Electric Power Technology.* 46 55-62

[2] Qi L and Xuebing W. Real time data transmission and processing of Internet based gas pipeline UAV patrol 2025 *J. China New Communications*. 27 8-19

[3] Haitao C, Zezhao W, Hongbo S, Linyu K and Luyao W. Research on security risk and prevention mechanism of UAV patrol data in power industry 2025 *J. Henan Electric Power.* 3 54-60

[4] Yue X. Research on Meteorological Disaster Warning and Online Monitoring Network Method for Transmission Channels 2022 D. *North China Electric Power University*

[5] Xue Z, Yinwen X and Meng Z. Risk assessment and prediction of important transmission channels based on water wave algorithm factor analysis long short-term memory network 2022 *J. Modern Electric Power*, 39 278-286

[6] Jiaxin B, Fan W, and Wen Z. A continuous lightning trip warning method for tight transmission Channels driven by lightning location data 2022 *J. Power Grid Technology.* 46 1194-1205

[7] H. Li Q. Tan H. Xie 2021 *Monitoring surface deformation of transmission channels using time series ALOS-2 PALSAR data* Surveying and Spatial Geographic Information, vol. 44, no. 6, pp. 46-50+53

[8] Detao W, Hong H, and Min Y. Research on Disaster Risk Warning and Auxiliary Decision Making for Dense Channels 2021 *J. Electrical Technology.* 10 185-186+189

[9] Panming W, Chaowen W and Weng Z. Research on the evaluation method of continuous lightning strikes in important transmission channels 2019 *J. High Voltage Electrical Appliances*. 55 178-185

[10] Li L, Qing W, Wenwen Z and Dan Q. Research on Intelligent Risk Pre control Technology for Transmission Line Channels Based on Big Data 2020 *J. Electrical Measurement and Instrumentation.* 57 82-87

[11] Xinyan X, Zeng J and Liuxi W. Risk assessment of forest transmission channels based on the spatiotemporal characteristics of wildfires 2018 *J. Power System Protection and Control.* 46. 1-9

Acknowledgment

Supported by Project of State Grid Electric Power Space Technology Company Limited: Research on Data Storage and Management Technology Based on SGCC's Internal and External Cloud and Self built Disaster Recovery Cluster Environment (52950024000A)

DynMixMFusion: dynamic mixture of experts multimodal fusion model

Yong Liu*, Zifeng Zhang, Jian Zhu, Jingya Li, Yingqi Zhang

State Grid Information & Telecommunication Group Company Ltd, Beijing Branch, Beijing, China

Corresponding email: 157725403@qq.com

Abstract: This paper proposes the Dynamic Mixed expert Multimodal Fusion model, aimed at improving the efficiency of automated compliance detection for operational behaviors in the inspection and operation of power distribution networks. The model consists of two parts, the MFusion model and the DynMix model. Firstly, the paper designs the MFusion model, introducing the Cross Feature Block and the Concatenation Feature Block to enhance the fusion of different modalities, improve detection accuracy, and increase the efficiency of compliance checks. Experiments show that in the semantic segmentation task, the model demonstrated a 1.8% improvement in mIoU, and in the sentiment analysis task, the accuracy reached 79.77% with MAE of 0.60, validating the strong multimodal fusion expression capability of the MFusion model. Secondly, the paper proposes the DynMix model, which employs a dynamic mixture of experts selection mechanism to choose different multimodal fusion strategies based on the difficulty of the task, significantly optimizing the allocation of computational resources. Ultimately, DynMixMFusion was tested and applied in the compliance detection of substations, achieving a classification accuracy of 87%, which improved the efficiency of accountability and traceability in power operation and inspection, providing a technical approach for real-time monitoring of compliance in power system operations.

1. Introduction

The operation and inspection of power distribution networks have strict requirements for the operational behavior of operators. If an operator performs incorrect actions, such as performing switching operations on the wrong objects, it can cause equipment damage or trigger power grid accidents, leading to catastrophic losses for the regional power grid[1]. To avoid the aforementioned situations, China power grid has operators' actions recorded by a recorder and verified on-site to minimize the possibility of incorrect actions being taken. However, in this scheme, the utilization rate of recorded videos is insufficient. When tracing accident responsibility, staff still need to play the entire recorded video to make judgments on the recorded actions, which greatly increases the cost of accident accountability when faced with a vast amount of video data. Therefore, there is an urgent need to propose an intelligent detection scheme to assist staff in making rapid judgments on the operational behaviors recorded in the videos. However, the video data recorded by the recorder often suffers from issues such as occlusion, lighting, and video quality, leading to blurry videos, unclear content, or even failure to capture the operational actions. Therefore, it is necessary to integrate various modalities of information, such as audio from the video, required text from operation tickets, and operational images, to fill in the gaps in different modalities. Traditional multimodal fusion techniques often directly integrate multiple modalities, and the model inference typically requires passing through a complex and computationally extensive multimodal fusion process. In existing multimodal fusion methods, the processing of images often relies on CNN[2] or ViT[3]. CNN-based methods have garnered significant attention due to their scalability and linear complexity. However, these methods are often limited by the size of the kernel and the small receptive field, which results in a poor ability to capture global information, handle long-range dependencies, and correlate image information with textual information. Methods based on the ViT model enhance the model's visual modeling capabilities by utilizing a global receptive field and dynamic weight mechanisms. However, due to their reliance on the Transformer model, these methods have high computational complexity and suffer from low detection efficiency.

At the same time, we have observed that there are some samples in the operation recording videos that are clearly shot and have good lighting, and they can still show relatively good detection effects without integrating voice modality features. Therefore, when dealing with simpler tasks, it is not necessary to consider multiple modalities comprehensively. For instance, for clear operation recording videos, it is sufficient to combine image and operational text without the need for video modality participation. For blurry operation recording videos, it is necessary not only to combine video images and operational requirement texts but also to consider the verbal communication between the operator and the supervisor to determine whether the operation is compliant, thereby providing timely alerts for non-compliant operations. However, existing multimodal fusion schemes often lack the ability to select different modality fusion strategies based on the difficulty of the task, leading to unnecessary waste of computational resources.

To address the aforementioned issues, this paper proposes the following solutions:

1. We propose the MFusion model, which constructs a Cross Feature Block (CroFB) and a Concatenation Feature Block (ConFB) to enhance the fusion capability of different modality features, thereby improving the detection accuracy for related tasks. In the relevant multimodal semantic segmentation tasks, the mIoU metric has been improved by 1.8 and 0.9 respectively.

2. We propose the DynMix model, which combined with the MFusion model, enhances the detection efficiency of the model. In the sentiment analysis task, while reducing the computational load of the model, the accuracy of sentiment analysis still reaches 79.77%, with an MAE of 0.60.

3. We have applied the DynMixMFusion model proposed in this paper to the compliance detection of power operation and inspection at a certain substation, achieving a compliance classification accuracy of 87%, providing technical support for the detection of compliance in power operation and inspection as well as accountability for non-compliant operations.

2. Related Work

Object detection and sentiment analysis have been hot topics in multimodal fusion research in recent years[4]. For instance, Lin et.al. [5] addresses the weak deep feature interaction between different modalities and the difficulty in distinguishing high-level semantic relationships between modal features in sentiment analysis problems by proposing a modality binding learning framework, which enhances the model's ability to express connections between different modality data. Jin et.al. [6] addresses the issue of existing tensor-based multimodal fusion methods that do not make full use of fine-grained temporal features. A dual low-rank multimodal fusion technique is proposed, which associates features between all modalities at different time steps to form a high-order tensor. This tensor is then embedded into a low-dimensional matrix, and low-rank factorization is used to approximate the dual dimensions of the high-order tensor, effectively enhancing the model's understanding of modality temporal changes. Maiti et.al. [7] to address the issue that traditional fusion of 3D point clouds and images can destroy the features of 3D point clouds (by reducing dimensionality and superimposing image features), and fail to fully utilize the presence information in the point cloud, a late fusion approach is adopted. Additionally, cross-attention is used to fuse the extracted image features and point cloud features, directly integrating the image with the point cloud, thus eliminating the need for lossy preprocessing of the point cloud.

Despite the existence of a multitude of multimodal fusion models, the enhancement of multimodal fusion efficiency and the reduction of training costs remain a significant focus of attention. Li et.al. [8] addresses the high training and fine-tuning costs of multimodal models, which require a large amount of memory, by proposing a new form of modular multimodal fusion framework. This framework utilizes three types of interactive prompts—query prompts, query-context prompts, and fusion-context prompts—to dynamically learn different objectives in multimodal learning. Wang et.al. [9] addresses the issue that aggregation-type fusion tends to neglect the propagation of single-modality features, while alignment-type fusion, although retaining single-modality features, has poor cross-modal fusion effects. It is challenging to both integrate the common information from different modalities and retain the specific patterns of each modality. To tackle this, the literature proposes a method for dynamically exchanging channels between different modality sub-networks, achieving effective cross-modal fusion while preserving single-modality features.

Despite a plethora of research findings, there remains a delicate balance between the complexity of model inference and the precision of model detection, much like the paradoxical desire to both keep and consume one's cake simultaneously. The pursuit of enhancing model detection efficiency without compromising the original accuracy poses a formidable challenge in contemporary research endeavors.

3. Methodology

This section will provide a detailed introduction to the proposed DynMixMFusion model for multimodal fusion. The DynMixMFusion model consists of two modules: the MFusion modality fusion module and the dynamic mixture of experts selection DynMix model. Firstly, we will introduce the MFusion model, which includes two modules: the Cross Feature Block (CroFB) and the Concatenation Feature Block (ConFB). By combining these two modules, different modality feature expressions can be fused, enhancing the accuracy of multimodal task analysis. Following this, we will outline the DynMix model, which is based on a dynamic mixture of experts selection mechanism. This mechanism selects different expert models based on the difficulty of the task, addressing the issue of low inference efficiency in static multimodal models.

3.1 MultiModal Fusion Model

The MFusion model often faces challenges with loosely associated features and difficulties in aligning correlated features across different modalities. Therefore, this paper proposes the MFusion multimodal feature fusion module, which is composed of the Cross Feature Block (CroFB) and the Concat Feature Block (ConFB). In the MFusion model, the CroFB module utilizes a cross-feature enhancement module to strengthen the interaction between features. The ConFB module, on the other hand, employs a long-term feature scanning module to achieve a more optimal feature fusion effect. If the features derived from the k modality are represented as $\mathcal{F}_k \in \mathbb{R}^{H \times W \times C}$, the fusion process can be expressed as follows:

$$\mathcal{F}_{fusion} = \mathrm{ConFB}\Big(\mathrm{CroFB}\big(\mathcal{F}_1, \mathrm{ConFB}\big(\mathrm{CroFB}\big(\mathcal{F}_2, \cdots\big)\big)\big)\Big) \tag{1}$$

If there are only two modalities, then equation (1) can be simplified to equation (2):

$$\mathcal{F}_{fusion} = \mathrm{ConFB}\big(\mathrm{CroFB}\big(\mathcal{F}_1, \mathcal{F}_2\big)\big) \tag{2}$$

In which $\mathrm{CroFB}(\mathcal{F}_1, \mathcal{F}_2)$ and ConFB represent individual modules, with specific details as follows:

The **CroFB** module, as shown in Figure 1, first processes the two input features through a linear layer and deep convolution, and then inputs them into the Cross Feature Enhancement (CFE) module. In the CFE module, linear projection layers are used to generate system matrices B, C and Δ from the input features. These system matrices are key to enhancing the model's context-aware capabilities, and by exchanging feature vectors between different modalities, the relationships between features are encoded into the model. Through this approach, the MFusion model is not only able to understand the features of individual modalities but also to associate the interactions and dependencies between different modality features. Inspired by the cross-attention mechanism [10], we have enhanced the ability of the cross-feature selection module to exchange information between features of different modalities. Specifically, the cross-feature exchange process can be represented as equation (3) to equation (7):

$$\overline{A}_1 = e^{\Delta_1 A_1}, \overline{A}_2 = e^{\Delta_2 A_2} \quad (3)$$

$$\overline{B}_1 = \Delta_1 B_1, \overline{B}_2 = \Delta_2 B_2 \quad (4)$$

$$h_1^t = \overline{A}_1 h_1^{t-1} + \overline{B}_1 x_1^t, h_2^t = \overline{A}_2 h_2^{t-1} + \overline{B}_2 x_2^t \quad (5)$$

$$y_1^t = C_2 h_1^t + D_1 x_1^t, y_2^t = C_1 h_2^t + D_2 x_2^t \quad (6)$$

$$y_{1/2} = [y_{1/2}^1, y_{1/2}^2, \cdots, y_{1/2}^l] \quad (7)$$

Where $x_{1/2}^t$, $h_{1/2}^t$, and $y_{1/2}^t$ represent the input, hidden state, and output of the model at time t, respectively, C_1 and C_2 are cross-modality matrices, as shown in Figure 1, which can span different modalities and share information between them.

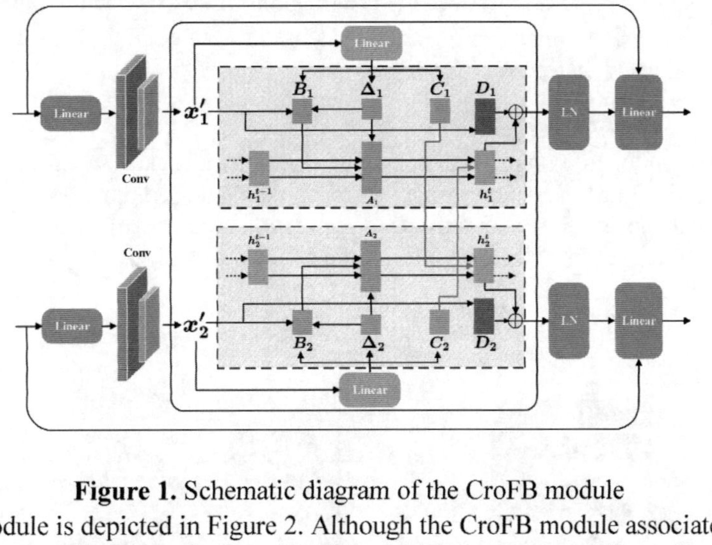

Figure 1. Schematic diagram of the CroFB module

The **ConFB** module is depicted in Figure 2. Although the CroFB module associates features of two modalities through cross-feature selection, thereby obtaining cross-modality enhanced features, the output of the module is still multiple unfused features that cannot be analyzed further. Therefore, to further extract fused features containing key information from multiple modalities, the ConFB module is employed to integrate the modality-enhanced features output by the CroFB module, thereby achieving multi-modality fusion.

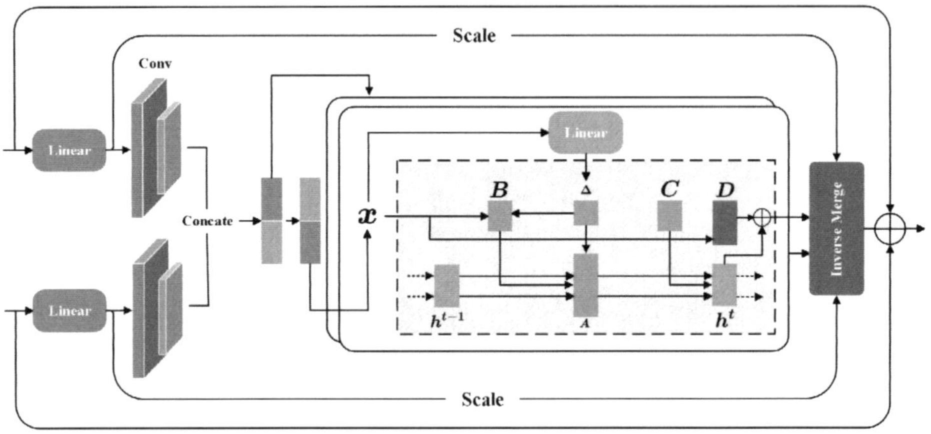

Figure 2. Schematic diagram of the ConFB module

Assuming the enhanced modal features from the CroFB module are $\widetilde{\mathcal{F}}_1 \in \mathbb{R}^{H \times W \times C}$ and $\widetilde{\mathcal{F}}_2 \in \mathbb{R}^{H \times W \times C}$, after undergoing linear transformation and convolution processing, both modal features are flattened and concatenated. This results in the concatenated feature vector $V_1 \in \mathbb{R}^{(2 \times H \times W) \times C}$, and subsequently, the reversed feature vector V_2 is generated. The two feature vectors are then input into the **L**ong-term **F**eature **S**canning (LFS) module to capture long-term dependencies within the sequence, resulting in $\tilde{V}_1$ and $\tilde{V}_2$. It should be noted that the structure of the LFS module is similar to that of the CFE module, implying that the execution process within the LFS module is akin to that of the CFE module. Afterwards, the reversed feature vector $\tilde{V}_2$ is added to $\tilde{V}_1$, and the resulting vector is decomposed into vectors $\overline{\mathcal{F}}_1$ and $\overline{\mathcal{F}}_2$. These two vectors are then multiplied by scaling parameters and merged along the channel dimension. Finally, a linear layer is used to scale the features into $\mathbb{R}^{H \times W \times C}$, yielding the final multi-modal fused feature vector.

3.2 Dynamic Mixture of Experts (DynMix) model

Through the aforementioned MFusion model, multimodal fusion can be achieved. However, as the number of modalities increases, the model's parameters and computational load will increase exponentially, which poses high demands on computational resources. Therefore, we propose the DynMix model to optimize the MFusion multi-modal fusion model, thereby accelerating the model's inference speed.

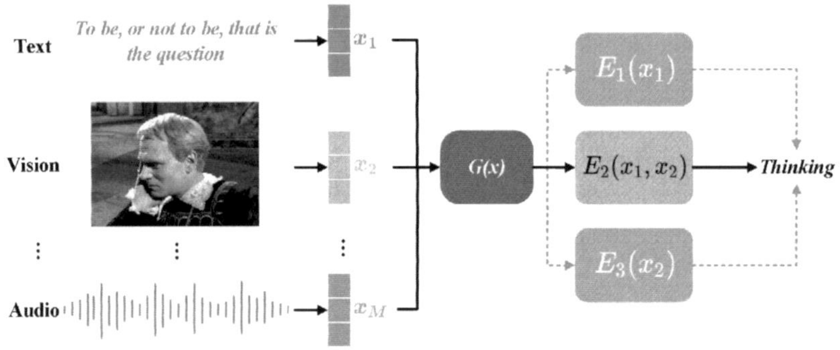

Figure 3. Dynamic Mixture of Experts Model

Journal of Physics: Conference Series **3036** (2025) 012004 doi:10.1088/1742-6596/3036/1/012004

For a complex task, the input data consists of M modalities, denoted by $x=(x_1,x_2,\cdots,x_M)$. To enhance the model's inference performance, this paper constructs several expert networks based on the Mixture of Experts (MoE) model. For instance, in the task of sentiment analysis, with $M=3$ modalities, one could design some expert networks, which can be denoted as $E_1(x_1)$, $E_2(x_2)$, $E_3(x_1,x_2)$ and additionally include $E_4(x_1,x_3)$, making a total of $C_3^1+C_3^2+C_3^3=7$ possible combinations. In practical applications, due to the excessive number of expert networks, the model's inference speed can become too slow. Therefore, only a subset of expert networks is often selected to be part of the Mixture of Experts model. Traditional Mixture of Experts (MoE) models often have all expert models perform task inference and then take a weighted sum of the inference results. However, the dynamic expert selection mechanism in DynMixMFusion does not execute all branches but instead selects a specific branch for inference. Specifically, as shown in Figure 3, let's assume that the number of selected expert networks is N. A dynamically constructed Mixture of Experts selection network $G(x)$ is used to determine which expert network to use for modality fusion and inference. The structure of $G(x)$ often needs to balance low computational cost and strong expressive power. In this paper, a Transformer network with 2 attention heads and a hidden layer size of 512 is used as the expert selection network. The input to the expert selection network is a global perception of the multimodal input x, and the output is an N-dimensional vector c representing which expert network is selected. Therefore, the final output of the task, as shown in equation (8), is:

$$\hat{y}=\sum_{i=1}^{N}c_iE_i(\mathrm{x}_i) \tag{8}$$

where $\hat{y}$ represents the inference result of the DynMixMFusion model, and $\mathbf{x}_i$ represents the modal input to the i-th expert network.

Since the decisions made by the expert $G(x)$ selection model are discrete and not directly differentiable, adjustments need to be made to the backpropagation network training method. During the model training phase, we use the Gumbel-softmax and reparameterization techniques to rewrite the selection vector c, expressed in the form shown in Equation (9):

$$\tilde{c}_i=\frac{e^{\frac{\log G(x)_i+b_i}{\tau}}}{\sum_{j=1}^{B}e^{\frac{\log G(x)_j+b_j}{\tau}}} \tag{9}$$

In which, $b_1,b_2,\cdots b_B$ is an independently sampled value from the Gumbel(0,1) distribution, and τ represents the temperature parameter. The larger the value of τ, the more uniform the distribution of $\tilde{c}$ becomes. Conversely, the smaller the value of τ, the more $\tilde{c}$ distribution resembles a categorical distribution. During forward propagation, the selection vector C is used to enhance the computational efficiency of the model. In the backward propagation, an approximation of the gradient $\nabla c=\nabla\tilde{c}$ is employed. Additionally, in order to enable efficient model inference and reduce the computational complexity of the model, we have added a computational resource cost function $\mathrm{C}(E_i)$ to the training objective on top of the existing model loss function. This function serves as the computational cost for selecting the expert network E_i. Given that the size of each expert network model can be statistically determined, the computational resource cost function can be predetermined before the training of the Mixture of Experts (MoE) selection model and is often set as a constant term. Therefore, the final training objective function is shown in Equation (10):

$$\mathcal{L}_{mix} = \mathcal{L} + \lambda \sum_{i=1}^{B} c_i C(E_i) \tag{10}$$

In which $\mathcal{L}$ represents the loss function for a specific task, such as the mean squared error between the model's predicted results and the actual values in a time series forecasting problem; λ serves as a hyperparameter that controls whether the training outcome is more inclined towards improving the model's performance (when λ is smaller) or more inclined towards reducing the model's inference complexity (when λ is larger).

4. Experimental Analysis

4.1 Description

To verify that our model can integrate multiple modalities of data and enhance semantic segmentation performance, we first assessed the model's performance on RGB-T and RGB-D datasets. By comparing the impact of different multimodal fusion models on segmentation performance, we demonstrated the effectiveness of the proposed multimodal fusion model in semantic segmentation tasks and showcased its broad applicability and efficient feature integration capabilities. Furthermore, we conducted sentiment analysis on the CMU-MOSEI sentiment analysis dataset using images, text, and audio to test the contribution of the dynamic mixture of expert selection model in improving analysis efficiency. Ultimately, we applied the model to real-world scenarios to evaluate its performance in practical applications.

4.1.1 Dataset

In this experiment, the datasets used and their corresponding data preprocessing operations are as follows:

1) **CMU-MOSEI dataset** [11] for the sentiment analysis task, monologue video clips from YouTube were collected, totaling 3,228 videos. Each video was segmented into short clips of about ten seconds, and each clip was accompanied by seven emotional ratings ranging from -3 to +3. During the training and validation of the model, we used 16,265 data points for training and 1,869 data points for model validation. The remaining data was reserved as a test set to evaluate the model's performance.

2) **MFNet dataset** [12] collected object images and corresponding thermal images from autonomous driving scenarios, including 820 daytime RGB-thermal images and 749 nighttime RGB-thermal images. To facilitate subsequent model training and evaluation, we extracted half of the data from these daytime and nighttime RGB-T images as the training set, a quarter as the validation set, and the remaining quarter as the test set.

3) **NYU Depth V2 dataset** [13] consists of a series of indoor scene video frames, totaling 464 different indoor environments. For training and testing, we selected 1449 annotated RGB-depth images from the dataset. These images were allocated to the training set and the test set, with 55% of the images used for model training and 45% for model evaluation.

4.1.2 Training Procedure for the Experiment:

The experimental training environment for this test is the Ubuntu 20.04 operating system, with 64G of system memory, including 1 CPU processor, 1 A800 GPU, corresponding to CUDA version 11.8, with 500 training epochs, a batch size of 8 per epoch, a learning rate of 5e-5, and a weight decay of 0.01.

4.2 The analysis of the results from the ablation study

In the analysis of the ablation study results for the multimodal fusion model, we first compared the mIoU performance of semantic recognition with and without multimodal fusion, as well as under different multimodal fusion techniques, on the MFNet dataset and the NYU Depth V2 dataset. The formula for calculating the mIoU metric is shown in Equation (11).

$$mIoU = \frac{1}{k+1}\sum_{i=0}^{k}\frac{TP}{FN+FP+TP} \tag{11}$$

The experimental results are shown in the following table:

Table 1. Performance of Semantic Segmentation Tasks on the MFNet Dataset

Method	Params (M)	FLOPs (G)	Car	Person	Bike	Curve	Stop	Guardrail	Cone	Bump	mIoU
RTFNet[14]	246	185	87.1	70.6	62.4	45.6	29.6	0.0	29.3	55.7	53.1
PSTNet[15]	106	123	76.5	52.3	55.6	29.9	24.9	**15.3**	39.2	45.2	48.1
CMX[16]	140	134	89.8	74.9	64.2	50.4	35.4	8.6	54.0	60.4	59.4
CMXNeXt[17]	120	132	91.3	75.1	67.3	**50.5**	40.4	8.9	53.1	52.3	59.4
Proposed	70	139	**91.4**	**75.7**	**67.9**	49.5	**41.9**	9.5	**55.0**	**60.4**	**61.2**

Table 2. Performance of Semantic Segmentation Tasks on the NYU Depth v2 Dataset

Method	Params(M)	mIoU
ACNet[18]	117	48.1
SGNet[19]	65	51.2
MultiMAE[20]	95	56.3
InvPT[21]	423	53.5
Proposed	70	**57.2**

Table 1 presents the results of various semantic segmentation tasks performed by different models on the MFNet dataset, while also comparing the size and computational complexity of the models. The results indicate that our model can achieve better performance than other benchmark methods with a lower number of parameters and floating-point operations. Additionally, our base model improved the mIoU by 1.8% compared to the second-best model. In Table 2, we further compared our model with a variety of other methods targeted at RGB-D scenarios to test its generalization capabilities. The comparison results show that our model also achieves good experimental results in the RGB-D semantic segmentation task, with a detection performance improvement of 0.9% over the second-best model.

Figure 4's semantic segmentation visualization effects demonstrate that our model can achieve precise segmentation results. This indicates that our approach can extract key features from different modalities, correlate and integrate different features, build bridges between various modal data, and thus obtain more comprehensive and accurate segmentation effects.

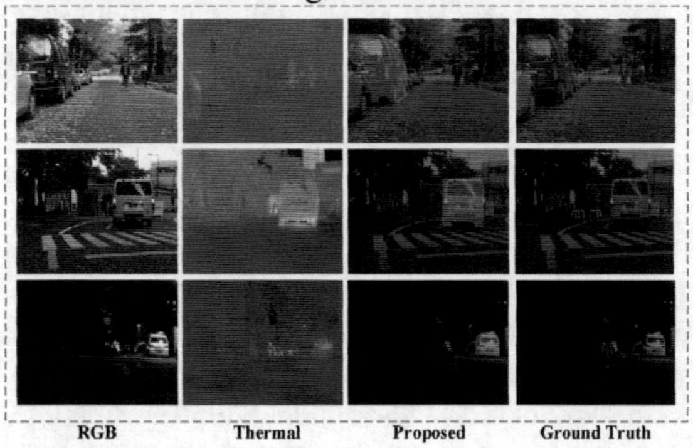

(a)

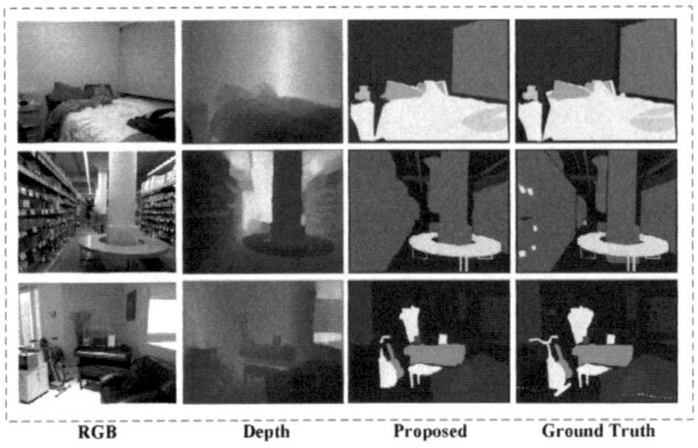

(b)

Figure 4. Model Segmentation Visualization Effects(a is the MFNet Dataset; b is the NYU Depth V2 Dataset)

4.3 Dynamic Mixture of Experts Model for Accelerating Inference: Experimental Comparison

Due to the increase in the number of modalities, the computational load of multimodal fusion models will become larger. Therefore, this experiment tests the sentiment analysis problem containing three modalities to verify whether the dynamic mixture of experts model proposed in this paper can improve detection efficiency. The test results are shown in Table 3.

Table 3. Sentiment Recognition Effects of Different Multimodal Fusion Strategies

Modality	Method	Acc	MAE	Operations(M)
unimodal	Video Network	69.06	0.81	123
	Audio Network	67.67	0.83	123
	Text Network(E_1)	78.34	0.63	125
MultiModal(Static)	Early Fusion	78.46	0.65	314
	Late Fusion(E_2)	79.52	0.61	310
MultiModal(Dyn)	DynMixFusion-a	79.05	0.62	166
	DynMixFusion-b	79.72	0.63	255
	DynMixFusion-c	**79.77**	**0.60**	296

As shown in Table 3, in the sentiment analysis task, the textual modality performs the best. Therefore, we selected a unimodal network with text features as input as the expert network E1. As for the second expert network (E2) in the mixture of experts model, it is a late fusion network responsible for receiving input data from image, text, and audio modalities. Subsequently, the gating network will make selections through a linear layer. This gating network is responsible for receiving features merged from three modalities (image, text, audio) and deciding which expert network to activate based on the specific circumstances of the sample during the inference process. This design allows the model to be more flexible and efficient when dealing with different inputs. As shown in Table 3, we present three dynamic mixture of experts models trained with different λ values. Compared to the best-performing static network (i.e., the late fusion model), although its accuracy slightly decreased (by 0.47%), the DynMM-a model reduced computational load by 46.5%. With the gradual increase in model computational load, the sentiment analysis model with dynamic mixture of experts mechanism continuously improves in analysis accuracy and gradually reduces the mean absolute error, ultimately achieving more accurate predictive effects at a lower computational parameter volume, striking a balance between model parameter size and accuracy. This result proves that the dynamic mixture of experts selection mechanism can enhance the model's inference efficiency.

4.4 Real-world Scenario Application Testing

This model was tested for its application in the task of compliance detection for power switchgear operations. By integrating multiple modalities such as text from operation tickets, video image data from operation recorders, and voice commands from operators, the model detects whether the operator's actions are compliant. The actual application scenario detection results show that the model can achieve a compliance detection accuracy of 87%. Although the accuracy is high, due to the strict regulatory training that switchgear operators undergo, the probability of non-compliant operations is relatively low, leading to an imbalanced data distribution. This could potentially result in overfitting, necessitating further data collection, annotation, and research. For instance, data could be collected during training sessions to achieve a more balanced data distribution.

5. Conclusion

In this study, addressing the issue of compliance inspection for power construction personnel's operation and maintenance, which involves multiple modalities of data and is difficult to achieve efficient compliance inspection, a dynamic hybrid expert multimodal fusion model is proposed. This model enhances the efficiency of multimodal fusion while maintaining the effectiveness of multimodal fusion detection as much as possible. Firstly, to enhance the multimodal fusion capability, expand the model's receptive field, and improve the model's detection capabilities. By stacking the CroFB and ConFB module, the MFusion multimodal feature fusion model is constructed. Subsequently, addressing the issue of redundant inference operations in the detection efficiency of static multimodal fusion models, we proposed a DynMix dynamic hybrid expert module. This module can integrate expert models from different modality combinations and select the appropriate experts for model inference based on different problems, thereby improving the model's inference efficiency and reducing the time required for model detection. Finally, we applied the DynMixMFusion model to the power operation and inspection scenario, achieving an accuracy of 87% in compliance detection, realizing the automated intelligent detection of power operation and inspection operations, and providing technical support for the detection of non-compliant operations and the accountability of non-compliant behaviors in power operation and inspection. Despite these achievements, there are still issues such as unbalanced sample distribution, inability to achieve real-time monitoring, and high false positive rates. In the future, we will collect more relevant data for the power scenario and continuously iterate and optimize the model based on this data, ultimately achieving online detection of non-compliant operations and ensuring the stable operation of new types of power systems.

References

[1] Lin C, Xu Q. 2023 *IEEE Transactions on Power Delivery*, **38**(3): 2022-33.

[2] Xie W, Wang C, Lin Z, et al. 2022 *Computerized Medical Imaging and Graphics*, **102**: 102128.

[3] Wang Y, Chen X, Cao L, et al. 2022 *proceedings of the 2022 IEEE/CVF Conference on Computer Vision and Pattern Recognition (CVPR)*, 18-24.

[4] Gan C, Fu X, Feng Q, et al. 2024 *Expert Systems with Applications*, **242**: 122731.

[5] Lin H, Zhang P, Ling J, et al. 2023 *Information Processing & Management*, **60**(2): 103229.

[6] Jin T, Huang S, Li Y, et al. 2020 *Association for Computational Linguistics*.

[7] Maiti A, Elberink S O, Vosselman G. 2023 *Proceedings of the 2023 IEEE/CVF Conference on Computer Vision and Pattern Recognition Workshops (CVPRW)*, 17-24.

[8] Li Y, Quan R, Zhu L, et al. 2023 *Proceedings of the 2023 IEEE/CVF Conference on Computer Vision and Pattern Recognition (CVPR)*, 17-24.

[9] Wang Y, Huang W, Sun F, et al. 2020 *Proceedings of the 34th International Conference on Neural Information Processing Systems.* Vancouver, BC, Canada **2020**: 406

[10] Chen C F R, Fan Q, Panda R. 2021 P*roceedings of the 2021 IEEE/CVF International Conference on Computer Vision (ICCV)*, 10-17.

[11] Bagher Zadeh A, Liang P P, Poria S, et al. 2018 *Proceedings of the 56th annual meeting of the association for computational linguistics (volume 1: Long papers)*. Melbourne, Australia, 2236-2246.

[12] Ha Q, Watanabe K, Karasawa T, et al. 2017 *2017 IEEE/RSJ international conference on intelligent robots and systems (IROS)*. **2017**: 5108-5115.

[13] SILBERMAN N, HOIEM D, KOHLI P, et al.2012 *Computer vision – ECCV 2012*. Berlin, Heidelberg: Springer Berlin Heidelberg, pp: 746-760.
[14] Sun Y, Zuo W, Liu M. 2019 *IEEE Robotics and Automation Letters*, 4(3): 2576-83.
[15] Shivakumar S S, Rodrigues N, Zhou A, et al. 2020 *2020 IEEE international conference on robotics and automation (ICRA)*. 2020: 9441-9447.
[16] Zhang J, Liu H, Yang K, et al. 2023 *IEEE Transactions on Intelligent Transportation Systems*, **24**(12): 14679-14694.
[17] Zhang J, Liu R, Shi H, et al. 2023 *2023 IEEE/CVF conference on computer vision and pattern recognition (CVPR)*. **2023**: 1136-1147.
[18] Hu X, Yang K, Fei L, et al. 2019 *2019 IEEE international conference on image processing (ICIP)*. **2019**: 1440-1444.
[19] Chen L Z, Lin Z, Wang Z, et al. 2021 *IEEE Transactions on Image Processing*, **30**: 2313-2324.
[20] Bachmann R, Mizrahi D, Atanov A, et al. 2022 *Computer vision – ECCV 2022* 348-367.
[21] YE H, XU D. 2022 *Computer vision – ECCV 2022* 514-530.

Optimization of Microgrid Operation Strategies Based on Carbon Asset Graph Data

Sen Pan*, Junfeng Qiao, Hongbin Qiu, Shiwen Dong

China electric power research institute Co. Ltd, Nanjing, China 210003

*yingzi108@126.com

Abstract. This study aims to minimize operational costs and carbon emission by systematically considering load constraints, charging and discharging constraints of energy storage equipment and carbon quota limitations. A dual layer optimization model for microgrid carbon asset operation and economic benefits is formulated. Considering carbon asset operation into outer planning model, the operation strategy for microgrid is optimized by utilizing an improved Grey Wolf algorithm. The inner demand response model improves users' electricity consumption behavior through incentive measures. Then, the optimized load curve improves the utilization rate of new energy and it is input to the outer model. Interactive iterations between the inner and outer dual layer models are conducted until the optimal configuration and load curve of each device are reached..

1. Introduction

Under the context of "carbon peak and carbon neutrality," the operation and management of carbon assets are closely tied to corporate value growth and have become urgent issues that businesses need to address. Chen et al. [1] proposed establishing a corporate carbon asset management system through three dimensions: carbon resource management, carbon emission reduction technology management, and carbon trading and carbon sink management, in order to devise a comprehensive carbon asset operation management pathway. Among these, carbon emission reduction technology management is a crucial part of the carbon asset management system. By adopting proactive emission reduction measures and optimizing carbon asset allocation, companies can achieve greater value enhancement and a competitive edge.

Microgrids play a crucial role in energy upgrading and transformation, as well as in promoting green and low-carbon development. They are a powerful tool for achieving the "dual carbon" goals. References[2-3] have explored the economic efficiency of microgrid operations, establishing optimization models with the objective of minimizing operating costs, thus deriving optimal economic operation strategies. However, the output and load of renewable energy sources are unpredictable and volatile, significantly affecting the stability of microgrid power supply. The energy transfer and power support capabilities of energy storage systems can effectively mitigate this issue[4]. The above studies focus on the economic benefits of microgrid operation and the stability of power supply, but do not consider reducing the carbon emissions of the system.

As a means of carbon reduction on the supply side, the carbon trading mechanism plays a key role in promoting the low-carbon operation of microgrids. To reduce carbon emissions during microgrid operation, References [5-6] integrated carbon trading costs into the objective function of microgrid optimization. Furthermore, consumer behavior plays a significant role in the economic and low-carbon performance of microgrids. Demand response (DR) is a strategy used to regulate electricity consumption by encouraging users to adjust their behavior through price signals or incentives. This approach helps to flatten demand peaks and fill consumption valleys, thereby enhancing the microgrid's ability to absorb renewable energy sources such as wind and solar power. As a result, DR not only improves the microgrid's efficiency but also contributes to carbon reduction[7].

Few studies on microgrid operation simultaneously address the integrated mechanisms of carbon asset management and demand response. In reality, effectively combining these strategies can not only help power companies achieve emission reduction targets but also enhance renewable energy utilization, reduce dependence on energy storage systems, lower operating costs, and maximize economic benefits. This paper focuses on a microgrid equipped with diverse resources, including photovoltaic systems, wind turbines, diesel generators, and energy storage devices. With substantial carbon reduction potential and high demands for dispatch and operational management, this microgrid serves as the basis for developing a two-layer optimization model that integrates carbon asset management and economic benefit strategies. The outer optimization layer addresses carbon asset management by optimizing equipment configuration under constraints such as load demand, energy storage charging/discharging, and carbon quotas. An improved Grey Wolf algorithm is employed to enhance the system's low-carbon performance and economic efficiency. The inner optimization layer focuses on demand response, incentivizing users to adjust their electricity consumption patterns. The two layers interact iteratively, refining solutions until optimal performance is achieved. A case study is conducted to show the effectiveness of the operational management strategies.

2. Carbon asset data

Carbon asset data typically includes energy data, carbon emission data, economic data, and social data. There are issues in the current aggregation, integration, and management of carbon asset data[8-10], such as poor data processing synergy, low data connectivity efficiency, and weak dynamic expansion capabilities of models, which create inconveniences for analysis applications based on carbon asset data [11]. By defining the structural graph of carbon asset data using graph theory, we aim to construct a carbon asset graph data model that transforms relational carbon asset data into a graph data structure, thereby enhancing the analytical efficiency of carbon asset data [12]. All business data tables are extracted from the relational model of carbon asset operation data, and nodes are created by transforming the entities represented in the data tables, where the primary key fields serve as the identifiers for the nodes. The columns corresponding to the rows of the data table are transformed into attributes of the nodes and attached to them to form the nodes. Additionally, all primary and foreign key relationships and associated entity tables are extracted from the relational model of carbon asset operation data, and their linking relationships are analyzed. From the collection of nodes, corresponding starting and ending nodes are identified based on the linking relationships, which are used to connect the nodes and form edges, including relationships such as belonging to, containing, owning, and dependency. Based on the aforementioned conversion methods from relational model elements to nodes and edges in the graph model, carbon asset operation data is integrated, facilitating the transformation of relational model elements into graph model elements, and thereby creating a collection of nodes and edges that provides a foundation for the subsequent construction of the carbon asset operation data graph model.

3. Microgrid structure model

The microgrid consists of photovoltaic arrays, wind turbines, diesel generators, energy storage systems, and industrial loads. The system architecture includes a DC bus, which aggregates the output of the photovoltaic panels, and an AC bus, which connects to the main grid while integrating the outputs of the wind turbines and diesel generators, as well as the input from AC loads. The system layout is illustrated in Figure 1.

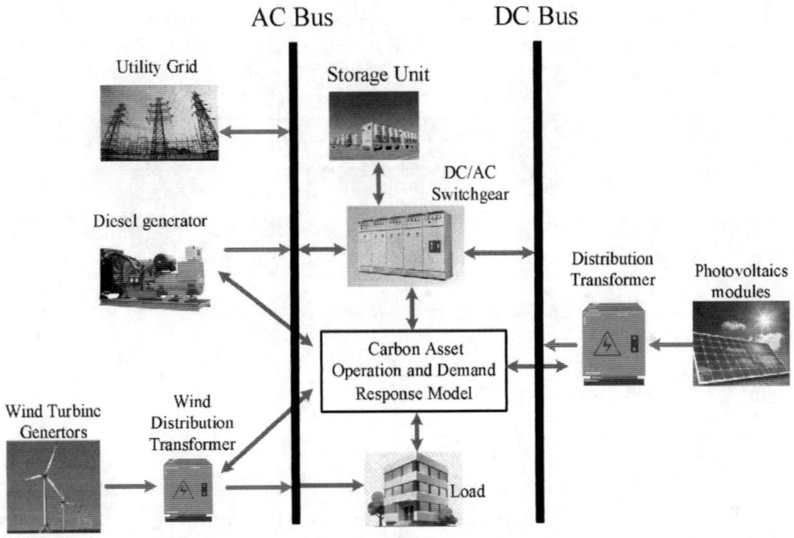

Figure 1. The microgrid system frame

3.1. Photovoltaic System model

A photovoltaic (PV) system is designed to generate usable electricity by harnessing solar energy through photovoltaics. The output energy of the PV module is given by (1):

$$P_{PV} = A_{PV} \cdot I_r \cdot \eta_{PV} \cdot t \tag{1}$$

Where , P_{PV} is output energy of PV modules, A_{PV} is the area of photovoltaic cells, $I_r$is irradiance, η_{PV} is photovoltaic cell efficiency, t is time.

3.2. Wind Turbine System model

A wind turbine is a device that converts the kinetic energy of wind into electrical energy.

The output energy of the wind turbine module is given by (2):

$$P_{WT} = \begin{cases} 0, v < v_i \\ P_{WTr} \cdot \dfrac{v^3 - v_i^3}{v_r^3 - v_i^3}, v_i \leq v \leq v_r \\ P_{WTr}, v_r \leq v \leq v_0 \end{cases} \tag{2}$$

Where, P_{WTr} is the rated output power of the wind turbine, v is wind speed, v_i is cut-in speed, v_r is rated wind speed, v_0 is cut-out speed.

3.3. Diesel generator system model

Diesel generators primarily serve as backup power sources. They are utilized as supplementary energy solutions to stabilize supply during periods when renewable energy fails to meet demand, helping to minimize energy fluctuations. The output energy of the diesel generator is given by (3):

$$E_{dg} = P_{dg} \cdot h \cdot d \tag{3}$$

where, P_{dg} is electricity generation , h is number of hours of electricity generation per day, d is number of hours.

3.4. Energy Storage System

The output of photovoltaic and wind power, as well as load demand, are inherently variable. An energy storage system plays a critical role in mitigating these fluctuations, enhancing the stability of the power supply. When the output from renewable energy sources exceeds load demand, the system stores the surplus energy until it reaches its rated capacity. Conversely, when demand surpasses supply, the system discharges the stored energy until it is fully depleted. Energy storage modules typically utilize batteries, with their charging and discharging power defined as follows:

$$P_{ch}(t) = \frac{S_{soc}(t) - S_{soc}(t-1)(1-\tau)}{\eta_c \Delta t} \tag{4}$$

$$P_{dis}(t) = \frac{[S_{soc}(t-1)(1-\tau) - S_{soc}(t)]\eta_d}{\Delta t} \tag{5}$$

where, τ is energy storage loss coefficient, η_c is battery charging efficiency, η_d is battery discharge efficiency, $S_{soc}(t)$ is the capacity of the battery at time t:

$$S_{soc}(t) = S_{soc}(t-1) + \eta_c \cdot P_{ch}(t-1) - \frac{P_{dis}(t-1)}{\eta_d} \tag{6}$$

4. Demand-Based Carbon Asset Dispatch Model

4.1. Microgrid operation model

Given that the load under study is an industrial load, the demand remains relatively stable throughout the day, with a distinct difference in demand between day and night. A 24-hour day can be divided into two periods: from 6:00 am to 6:00 pm (daytime) and from 6:00 pm to 6:00 am (nighttime). Let the load during the day be represented as $P_{load}^{D}(t)$ and the load at night as $P_{load}^{N}(t)$. The average output power of a single photovoltaic module is denoted as $P_{pv}(t)$, while the average output power of a single wind turbine is $P_{WT}(t)$, and the output power of a single diesel generator is P_{dg} .When the output of the power generation system exceeds the load, the excess energy is either stored in the battery or sold to the grid. Conversely, when the power generation output is less than the load, the battery will release stored energy to compensate for the shortfall, or electricity will be purchased from the grid. Furthermore, demand response can be utilized to encourage users to modify their electricity consumption patterns, thereby enhancing the microgrid's ability to integrate and accommodate renewable energy sources.

Decision variables of microgrid operation optimization model: number of photovoltaic power generation equipment in operation N_{pv}、 number of wind turbines in operation N_{wind}、 Current number of diesel generators in operation during daytime hours N_{dg}^{D}、 Number of diesel generators operating at night N_{dg}^{N}、 number of energy storage devices N_{ess}。 C_{pv} is photovoltaic power generation cost、 C_{wind} is wind power cost 、 C_{dg} is diesel power generation cost, C_{setup} is operating startup cost for a single generator unit, C_{ess} is operating cost of energy storage system.

4.2. Microgrid Carbon Asset Model

Under the carbon trading mechanism, carbon emissions, as a tradable commodity, become an asset of the enterprise, namely carbon assets. Enterprises can buy or sell carbon emissions depending on whether their actual annual emissions exceed their allocated carbon quotas. In this article, the microgrid's carbon emission sources primarily include emissions from diesel generators and purchased electricity. Assume that the carbon quota of the system is D_c, σ_{dg} is the carbon emission intensity per unit of power generation for diesel generators. σ_{buy} is the carbon emissions from purchasing electricity generated by thermal power plants, σ_{dg} is the carbon emissions from purchasing electricity generated by thermal power plants. λ is carbon pricing, average daily carbon asset operating costs C_c:

$$C_{co_2} = \lambda \left[\frac{\sigma_{dg} E_{dg} \left(N_{dg}^{D} + N_{dg}^{N} \right) \times 12 + \sigma_{buy} Q_{buy}}{1000} - \frac{D_c}{365} \right] \tag{7}$$

Where, Q_{buy} is average daily carbon asset operating cost.

4.3. Demand Response Model

The goal of demand response optimization is to minimize the total difference between renewable energy generation and load demand at each time point throughout the 24-hour period. Through pricing or incentive mechanisms, users are encouraged to adjust their electricity consumption behavior in line with the renewable energy output curve, thereby achieving peak shaving and valley filling.

The optimization objective function of demand response is:

$$min\sum_{t=1}^{24}\left|P_{pv}(t)+P_{WT}(t)-P_{load}^{'}(t)\right|\Delta t \tag{8}$$

where, $P_{load}^{'}(t)=P_{load}(t)+\Delta P_{load}(t)$ is new load power after demand response, $\Delta P_{load}(t)$ is the load transfer at time t, $\Delta P_{load}(t)>0$ indicates load transfer in, $\Delta P_{load}(t)<0$ indicates load transfer out. The total load transfer throughout the day must comply with the balance constraint, while the load transfer at each moment must meet the transfer constraint:

$$\sum_{t=1}^{24}\Delta P_{load}(t)\Delta t=0 \tag{9}$$

$$-\Delta P_{load}^{max}(t)\le\Delta P_{load}(t)\le\Delta P_{load}^{max}(t) \tag{10}$$

Where, ΔP_{load}^{max} is the maximum allowable load transfer at time t.

5. Carbon asset operation-economic benefit two-layer optimization model

A two-layer optimization model is developed for the operation of microgrid carbon assets and economic benefits. The outer layer planning model (OLPM) optimizes the configuration and operation of microgrid devices, while the inner planning layer model (ILPM) focuses on optimizing demand response. The results from the OLPM are passed to the ILPM for further optimization, and the optimized load curve from the ILPM is then fed back to the outer OLPM. This interaction continues iteratively between the layers until a convergent equilibrium is reached, where both economic and carbon-related objectives are balanced.

5.1. Outer layer planning model (OLPM)

The goal of the OLPM is to minimize the daily operating cost of the microgrid C_{run}, including the cost of photovoltaic power generation C_{run}^{pv}、 the cost of wind power generation C_{run}^{wind}、 the cost of diesel generation C_{run}^{dg}、 the cost of energy storage system C_{ess}^{run}、 purchase and sale electricity cost $C_{run}^{buy-sell}$ and carbon asset operational cost C_c, the calculation formulas are as follows::

$$C_{run}^{pv}=N_{pv}\cdot C_{pv}\cdot\sum_{t=1}^{24}P_{pv}(t)\Delta t \tag{11}$$

$$C_{run}^{wind}=N_{wind}\cdot C_{wind}\cdot\sum_{t=1}^{24}P_{WT}(t)\Delta t \tag{12}$$

$$\begin{aligned}C_{run}^{dg}= &\underbrace{12\cdot N_{dg}^{D}\cdot C_{dg}\cdot P_{dg}}_{Daytime\ power\ generation\ cost}+\underbrace{12\cdot N_{dg}^{N}\cdot C_{dg}\cdot P_{dg}}_{Nighttime\ power\ generation\ cost}\\ &+\underbrace{C_{setup}\cdot(N_{dg}^{D}-N_{dg}^{N})\cdot\chi\{N_{dg}^{D}>N_{dg}^{N}\}}_{The\ startup\ and\ shutdown\ cost\ of\ power\ generation}\\ &=12\cdot\left(N_{dg}^{D}+N_{dg}^{N}\right)\cdot C_{dg}\cdot P_{dg}+C_{setup}\cdot\left(N_{dg}^{D}-N_{dg}^{N}\right)\cdot\chi\{N_{dg}^{D}>N_{dg}^{N}\}\end{aligned} \tag{13}$$

where： $C_{setup} \cdot (N_{thermal}^{D} - N_{thermal}^{N}) \cdot \chi\{N_{thermal}^{D} > N_{thermal}^{N}\}$ indicates the startup cost of a diesel generator set that is shut down at night and starts running during the day, $\chi\{A\}$ *is* the indicator function $\chi\{A\} = \begin{cases} 1, & A\ is\ ture \\ 0, & A\ is\ fasle \end{cases}$。

$$C_{ess}^{run} = N_{ess} \cdot c_{ess} \cdot \left[\eta_c \cdot \sum_{t=1}^{24} P_{ch}(t)\Delta t + \frac{\sum_{t=1}^{24} P_{dis}(t)\Delta t}{\eta_d} \right] \tag{14}$$

where, c_{ess} is the unit energy operating cost of a single energy storage device (Unit: yuan/kw).

$$C_{run}^{buy-sell} = \sum_{t=1}^{24} [C_{buy}Q_{buy}(t) - C_{sell}Q_{sell}(t)]\Delta t \tag{15}$$

Where, C_{buy}、C_{sell} are import and export electricity cost respectively (Unit: yuan /kwh); $Q_{buy}(t)$、$Q_{sell}(t)$ are the power import and export at time t.

The objective function of the OLPM is:

$$min\, C_{run} = C_{run}^{pv} + C_{run}^{wind} + C_{run}^{dg} + C_{ess}^{run} + C_{run}^{buy-sell} + C_{co_2} \tag{16}$$

The constraints are as follows:

(1) for generator units and energy storage systems:

$$0 \le N_{pv} \le N_{pv}^{max} \tag{17}$$

$$0 \le N_{wind} \le N_{wind}^{max} \tag{18}$$

$$0 \le N_{dg} \le N_{dg}^{max} \tag{19}$$

$$0 \le N_{ess} \le N_{ess}^{max} \tag{20}$$

Where N_{pv}^{max}、N_{wind}^{max}、N_{dg}^{max}、N_{ess}^{max} are respectively the maximum available number of units for photovoltaic equipment, wind power equipment, diesel power generation equipment, and energy storage equipment.

(2) Charging and discharging power constraints and energy capacity constraints of energy storage equipment.

$$P_{ch}^{min} \le P_{ch}(t) \le P_{ch}^{max} \# \tag{21}$$

$$P_{dis}^{min} \le P_{dis}(t) \le P_{dis}^{max} \# \tag{22}$$

$$P_{ch}(t) \cdot P_{dis}(t) = 0 \# \tag{23}$$

$$S_{soc}^{min} \le S_{soc}(t) \le S_{soc}^{max} \# \tag{24}$$

Where $P_{ch}(t)$、$P_{dis}(t)$ represent the charging and discharging power at time t respectively. P_{ch}^{min}, P_{ch}^{max} represent the minimum and maximum charging power respectively, $P_{dis}^{min}, P_{dis}^{max}$ are the minimum and maximum discharging power respectively. , S_{soc}^{min}, S_{soc}^{max} are the minimum and maximum remaining capacities respectively. Equation (23) indicates that at any given moment, the energy storage device can only be in one state, either charging or discharging.

(3) Power Balance Constraint

$$\begin{aligned} &N_{pv} \cdot P_{pv}(t) + N_{wind} \cdot P_{WT}(t) + P_{dg} \cdot N_{dg}^{D} \\ &+ Q_{buy}(t) - Q_{sell}(t) + N_{ess} \cdot [P_{dis}(t) - P_{ch}(t)] = P_{load}^{D}(t) \quad t \in [6,18] \end{aligned} \tag{25}$$

$$\begin{aligned} &N_{pv} \cdot P_{pv}(t) + N_{wind} \cdot P_{WT}(t) + P_{dg} \cdot N_{dg}^{N} + Q_{buy}(t) - Q_{sell}(t) \\ &+ N_{ess} \cdot [P_{dis}(t) - P_{ch}(t)] = P_{load}^{N}(t), t \in [1,6] \cup [18,24] \end{aligned} \tag{26}$$

 doi:10.1088/1742-6596/3036/1/012005

5.2. Inner layer planning model (ILPM)

Inner layer planning model is:

$$min\sum_{t=1}^{24}\left|P_{pv}(t)+P_{WT}(t)-P_{load}^{'}(t)\right|\Delta t \tag{27}$$

$$s.t.\begin{cases} P_{load}^{'}(t)=P_{load}(t)+\Delta P_{load}(t) \\ \sum_{t=1}^{24}(t)\Delta P_{load}\Delta t=0 \\ -\Delta P_{load}^{max}(t)\leq \Delta P_{load}(t)\leq \Delta P_{load}^{max}(t) \\ C_r\sum_{t=1}^{24}P_{load}^{'}(t)-C_e\sum_{t=1}^{24}|\Delta P_{load}(t)|\Delta t\leq C_r\sum_{t=1}^{24}P_{load}(t) \end{cases} \tag{28}$$

The last constraint mentioned above ensures that the user's electricity cost after participating in demand response is less than that before participation, serving as an incentive. C_r is the standard electricity rate paid by the user, C_e is the unit demand response incentive.

5.3. The solution process of the two-layer planning model

The solution process is as follows:

(1) Input various operational parameters: load curve, parameters of each power generation equipment, energy storage equipment parameters, carbon asset operation parameters, and demand response parameters, etc.

(2) Optimize the outer-level planning model: The optimal equipment capacity configuration is obtained by minimizing the daily average total operating cost of the microgrid.

(3) Optimize the inner-level planning model: The optimization results from the OLPM are input into the ILPM to optimize and obtain the new load curve after implementing demand response, which is then fed back to OLPM.

(4) Determine whether the iterative convergence condition is satisfied: The results of two consecutive iterations remain approximately unchanged: $\left|C_{run}^{k+1}-C_{run}^{k}\right|/C_{run}^{k}\leq \varepsilon$

Where C_{run}^{k} is the result of the k-th iteration of the microgrid operating cost, ε is the iteration termination error is ε, set to 10^{-4} in this case. Repeat the above steps until the convergence condition is met, and output the optimal result.

The solving process of the model is shown in Figure. 2.

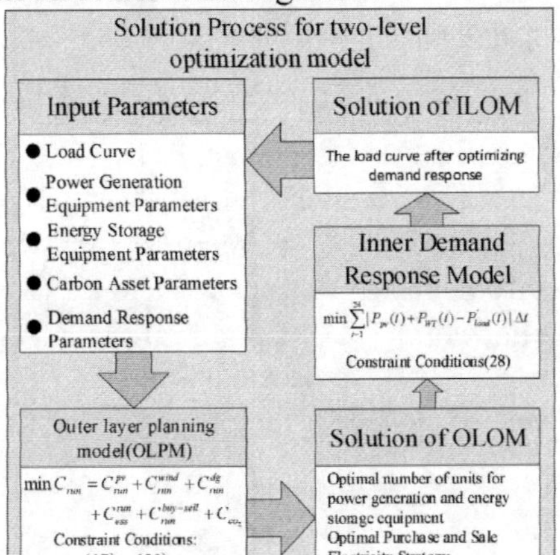

Figure 2. Low chart for solving the two-level optimization model

5.4. Improved Grey Wolf Optimization (GWO) Process

The Grey Wolf Algorithm is a swarm intelligence optimization algorithm that simulates the social leadership hierarchy and hunting behavior of grey wolves[10]. The basic Grey Wolf Algorithm is prone to falling into local optima in the later iterations due to a rapid decline in population diversity, leading to premature convergence. This paper improves the Grey Wolf Algorithm by incorporating a Logistic-Tent chaotic mapping, a nonlinear convergence factor, and a Cauchy mutation operator to enhance its global search capability, thereby providing a powerful tool for optimizing the outer-layer capacity planning model presented herein.

This paper employs the Logistic-Tent chaotic map during population initialization to enhance population diversity and ergodicity while accelerating convergence speed. The mapping equation is:

$$x_{n+1} = \begin{cases} [rx_n(1-x_n) + \frac{(4-r)x_n}{2}] \, mod\, 1, & x_n < 0.5 \\ [rx_n(1-x_n) + \frac{(4-r)(1-x_n)}{2}] \, mod\, 1, & 0.5 \le x_n \le 1 \end{cases} \tag{29}$$

where, n represents the current iteration count; r is a control parameter, set to r=0.1

This paper also introduces a nonlinear convergence factor that decreases slowly in the early iterations—thereby enhancing the algorithm's global search performance—and more rapidly in the later iterations to improve convergence. The expression for the nonlinear convergence factor is as follows:

$$a = a_0 + (a_1 - a_0)[1 - exp\,(-k\left(\frac{n}{N_{max}}\right))]^{-1} \tag{30}$$

where a_0 and a_1 correspond to the initial and terminal values of the convergence factor, respectively; $k \in [1,5]$ is the regulation coefficient; n is the current iteration count; N_{max} is the specified maximum iteration count. In order to further overcome the phenomenon of the algorithm's rapid decline in population diversity in the later stages leading to premature convergence, a Cauchy mutation operator is introduced here to maintain the diversity of the population during the iteration process, while also taking into account the convergence speed of the algorithm. The mutation operator is as follows:

$$X_g(n) = X_g(n) + \exp\left(-\frac{\lambda n}{N_{max}}\right) \cdot C(0,1) \tag{31}$$

where, $X_g(n)$ is the optimal solution for the n-th generation. ; $\lambda \in [30,100]$ is adjust parameter; $C(0,1)$ is standard Cauchy distributed random variable。

The improved Grey Wolf Optimization (GWO) process is as follows:

1) **Set Parameters**: Set population size, maximum number of iterations, and adjustment coefficients, and define the search space.
2) **Generate Initial Population**: Utilize the Logistic-Tent chaotic mapping to generate the initial grey wolf population within the search space, set n=1.
3) **Calculate and Rank Fitness Values**: For strategy $(N_{pv}, N_{wind}, N_{dg}^D, N_{dg}^N, N_{ess})$, the fitness value is calculated using Equation (16). The top three grey wolves with the best fitness values are selected, and their corresponding positions are denoted as X_α、X_β、X_δ.
4) **Check Termination Condition**: If $n > N_{max}$, the algorithm outputs the current best grey wolf individual and terminates; otherwise, proceed to Step (5).
5) **Update Grey Wolf Positions and Fitness Values**: Wolves in the population follow the top three grey wolves and gradually move toward the prey position. The position update formula is:

$$D_p = C \cdot X_p(n) - X(n) \tag{32}$$

$$X(n+1) = \frac{1}{3} \sum_{p \in \{\alpha,\beta,\delta\}} [X_p(n) - A \cdot D_p] \tag{33}$$

Journal of Physics: Conference Series **3036** (2025) 012005 doi:10.1088/1742-6596/3036/1/012005

Where D_p is the distance between the grey wolf and the prey; $X_p(n)$ is the position of the prey; $X(n)$ is the current position of the grey wolf. $C = 2r_1$, $A = (2r_1 - 1)a$, $r_1, r_2 \in random(0,1)$, a is the convergence factor, given by equation (30), its initial value is 2, and the termination value is 0. Recalculate the fitness value of the grey wolf individual.

6) **Conduct the Cauchy mutation operation:** Determine the current best wolf individual and apply the mutation operation according to equation (31) to generate a new wolf individual. Then, set $n = n+1$ and proceed to step (3).

6. Illustrative example

6.1. System parameters

The load distribution is shown in Figure.3. Tables 1 and 2 present the parameters for power generation equipment and energy storage devices, respectively. The selected location is in Lingang, Shanghai, at latitude 30.95861° N and longitude 121.89744° E. Based on historical meteorological information, the load curve and energy output curves are shown in Figures 3-5. Based on considerations of minimum and maximum load, related energy output, and available area, the recommended upper limits are: 200 for photovoltaic modules, 40 for wind turbines, 20 for diesel generators, and 20 for energy storage modules. The carbon quota is 2000 tons/year, with a carbon trading price of 80 yuan per ton. The electricity purchase price is 0.49 yuan per kWh, and the electricity sale price is 0.38 yuan per kWh.

Table 1. Power Generation Equipment Parameters

System parameters	Photovoltaic panels	Wind turbines	Diesel generators
Initial cost (CNY/kW)	0.3	0.046	0.16
Annual operation cost (CNY/kW)	0.14-0.28	0.017-0.030	0.04
Replacement Cost (CNY/kW)	--	80% of Initial Value	85% of Initial Value
Lifetime (years)	25	20	10
Carbon Emissions per kWh	0	0	0.78kg

Table 2. Energy Storage Equipment Parameters

Equitment parameters	η_c	η_d	τ	c_{ess} (CNY/kW)	P_{ch}^{min} (CNY/kW)	P_{ch}^{max} (MW)	P_{dis}^{min} (MW)	P_{dis}^{max} (MW)	S_{soc}^{min} (MW)	S_{soc}^{max} (MW)
Value	0.91	0.94	0.01	0.42	0.1	0.3	0.1	0.3	0.1	0.6

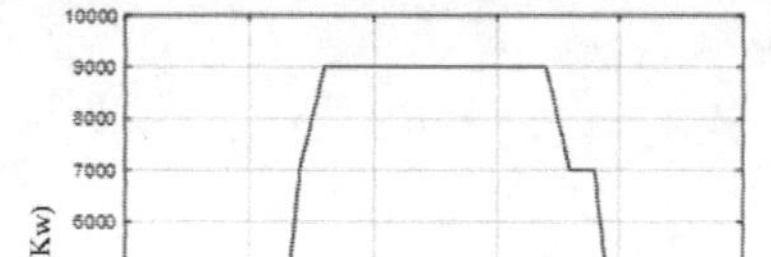

Figure 3. Daily average load curve

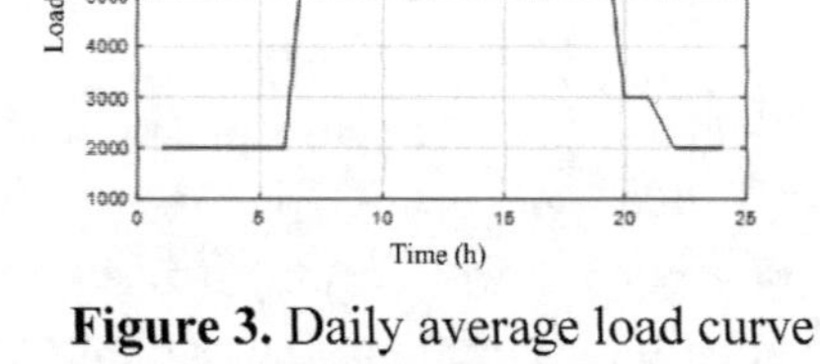

Figure 4. Daily average output power of PV

The function expression for the average output rate of the photovoltaic power generation module with respect to time, obtained by fitting based on historical average data, is as follows:

$$P_{PV}(t)=\begin{cases}0.01t^4-0.52t^3+8.39t^2\\ -49.33t+96.51,\ t\in[6,18]\\ 0, other\end{cases} \tag{34}$$

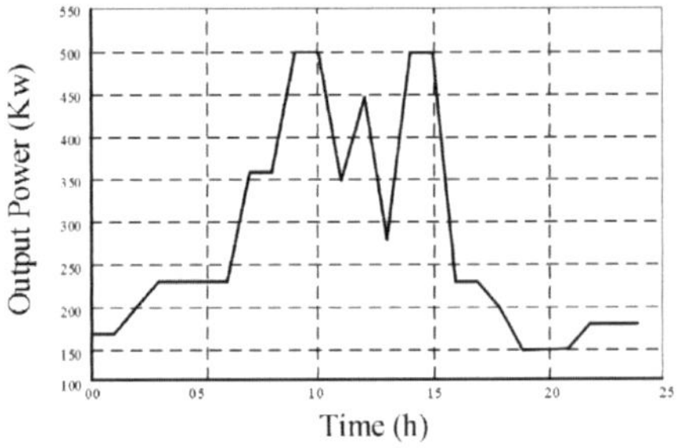

Figure 5. Daily average output power of WT

The piecewise function expression for the average output power of a single wind turbine at this location with respect to time, obtained by fitting based on historical average data, is as follows:

$$P_{WT}(t)=\begin{cases}-0.10t^5+2.31t^4-17.52t^3+51.66t^2\\ -33.54t+169.10, \quad t\in[1,10]\\ -52.92t^4+2756.26t^3-5357.01t^2\\ +460700.36t-1479000.97,\ t\in(10,15]\\ 0.08t^5-9.13t^4+382.87t^3-7954.32t^2\\ +81940.54t-334300.62, \quad t\in(15,24]\end{cases} \tag{35}$$

6.2. Optimization Results of the Bi-level Programming Model

The improved GWO is employed to optimize the outer-layer planning model, with a population size of 30, and a maximum of 100 iterations. The interaction iteration results of the inner and outer bi-level planning models are shown in Figure. 6. When the inner and outer iterations reach the fifth interaction, the error meets the iteration termination condition.

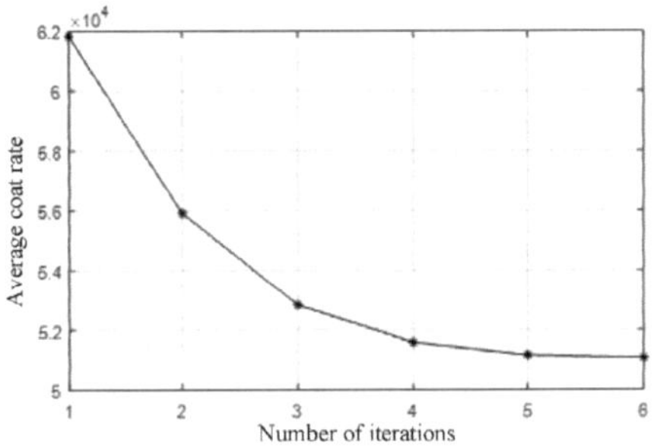

Figure 6. Iterative convergence curve

The optimal solution obtained after optimization is $N_{pv}=126$, $N_{wind}=14$, $N_{dg}^{D}=3$, $N_{dg}^{N}=0$, $N_{ess}=9$, which corresponds to the deployment of 126 photovoltaic (PV) generation modules, 14 wind turbines, operation of 3 diesel generators during the daytime with no operation at night, and the use of 9 energy storage modules to meet the daily average load after demand response. This configuration minimizes the total cost to $C_{total}=51062$ yuan.

7. Conclusion

This paper investigates the operational management strategy for microgrids. A dual-layer planning model is established with the objective of minimizing operating costs, taking into account the impacts of carbon quotas and carbon trading prices. The collaborative strategy for carbon asset management and demand response is optimized using an improved grey wolf algorithm. A case study of a microgrid is used to validate the effectiveness of the operational strategy. Due to the significant influence of weather and climate on renewable energy generation, energy storage modules help mitigate fluctuations in PV and wind power output, enhancing the overall stability of the grid. Demand response optimizes users' electricity consumption behaviors, further reducing unused energy stored in the system and decreasing the microgrid's dependence on the external grid for power supply. In future research, factors such as the randomness of distributed generation and fault transfer will be further considered, and the low-carbon and economic configuration of the microgrid will be optimized based on reliability assessment.

References

[1] Chen Jiangnin, Wang Lidon, Xia Wei, et al. Exploration of Carbon Asset Management Model under the "Dual Carbon" Goal. China Development Observation, 2021(16): 37-38+47.

[2] Sanseverino E R, Di Silvestre M L, Ippolito G, et al. An execution, monitoring and replanning approach for optimal energy management in microgrids. Energy, 2011, 36(5): 3429-3436.

[3] Li Xianbao, Zhang Ke. Research on optimal dispatching of microgrid based on improved harmony search algorithm. Journal of Northeast Electric Power University, 2022, 42(5): 83-89.

[4] Huang Xinghua1, Wu Han, CHEN Shichuan, et al. An Optimization Method for Energy Storage Configuration of Isolated Island Microgrid Considering New Energy Output. Electric Power, 2024, 57(12): 132-138.

[5] Lei Zhenhua, Lu Li, Liu Huan, et al. Optimal scheduling method for multi-microgrid system based on energy trading mechanism. Distribution & Utilization, 2024, 41(06): 36-46.

[6] Zhang Dongshun, Quan Hengli, Xie Hua, et al. Dispatching Strategy of Park-Level Integrated Energy System Considering Carbon Trading Mechanism and Hydrogen Blending Natural Gas. Electric Power, 2024, 41(06): 36-46.

[7] Huang Dongmei, Lü Jiaxin, Shi Shuai, et al. Optimal operation of island microgrid clusters considering demand response. Power System Protection and Control, 2024, 52(09): 88-98.

[8] LU Jun. Strategy for building a digital management system of carbon assets in group enterprises. Xinjiang Oil & Gas，2022，18(2)：10-15.

[9] Liu Y, Tian L, Sun H, et al. Option Pricing of Carbon Asset and Its Application in Digital Decision-making of Carbon Asset[J]. Applied Energy, 2022, 310: 118375.

[10] PIAO Zheyong, MENG Tao, ZHENG Wenyuan, et al. Research on Corporate Carbon Asset Value Assessment from the Perspective of Carbon Trading. Price: Theory and Practice, 2023(08): 81-86.

[11] Liu, Guangyi, Dai, Renchang, Lu, Yi, et al. Power Graph Computing Platform and Its Application in Energy Internet. Power System Technology, 2021, 45(6): 2051-2063.

[12] Huang, Quanlong, Huang, Yanxiang, Shao, Yingxia, et al. HybriG: An Efficient Attribute Graph Storage Architecture for Handling a Large Number of Duplicate Edges. Chinese Journal of Computers, 2018, 41(8): 1766-1779.

Acknowledgement

This work was supported by State Grid Corporation of China’s Science and Technology Project (5700-202312315A-1-1-ZN) which is 'Research and application of digital support technology for large-scale urban carbon asset operation services'.

Construction of electricity carbon emission prediction model based on multiple regression model

Zengwei Li[1], Jihong Song[2], Xiancheng Zhong[2], Hailong Li[2]*

[1]Information and Communication Company of State Grid Qinghai Electric Power, Qinghai, China
[2]State Grid Qinghai Electric Power Company, Qinghai, China

Corresponding email: lihl2019a@163.com

Abstract. The rapid development of the global economy has inevitably caused great harm to the natural environment. Electricity occupies an important share of the national economy and is characterized by high pollution and large carbon emissions (CEs). With the development of economy and the increase of greenhouse gas emissions, the global temperature generally rises and the environment is also seriously affected. In order to alleviate the environmental pollution problem and meet the needs of energy conservation and emission reduction in the construction power industry, this paper constructs a prediction model of electric power CEs, identifies the key factors affecting electric power CEs, and uses multiple regression model to study the influencing factors of CEs. The purpose is to put forward measures for energy conservation and emission reduction in electric power and reduce CEs in the construction industry. According to the analysis results of the model proposed in this paper, the correlation between CEs and power generation of thermal power generation, wind power generation, nuclear power generation, solar power generation and other power generation modes varies according to their characteristics. In the process of promoting sustainable development and protecting the environment, priority needs to be given to the use of low-CEs, efficient and energy-saving power generation methods.

1. Introduction

In order to reduce greenhouse gas emissions, CEs trading has become an important market mechanism. Under this mechanism, carbon emission(CE) rights become a tradable asset, and their value is affected by many factors, including market supply and demand, policy changes, technological progress, etc.

The emission of a large amount of greenhouse gases can lead to global temperature rise, causing adverse effects on terrestrial and marine ecosystems, leading to land drought, wetland degradation, sea level rise, and seawater acidification. These changes not only alter the distribution range of species, but also have negative effects on the interrelationships within and outside the ecosystem, and may even lead to the death of animals and plants, species variation, and a decrease in biodiversity. Therefore, taking measures to reduce greenhouse gas emissions and address climate change is currently crucial. As the main responsible party for global warming and environmental pollution, enterprises should take on more responsibility for carbon reduction. However, reducing the use of traditional fossil fuels on a large scale will lead to increased operating costs and financial risks, making it more difficult for companies to achieve net zero emissions targets. Some companies, in order to avoid rising costs, are unwilling to actively take emission reduction actions and ignore the carbon performance of the entire supply chain[1].

By improving investment policies, fiscal and taxation pricing policies, carbon markets and other market-oriented mechanisms, we can promote the process of carbon reduction. In terms of investment policies, we will establish a government led low-carbon investment and financing system, increase support for green and low-carbon investment projects, and promote green and low-carbon investment by market entities. In terms of fiscal and taxation pricing policies, increase financial investment in the development of green and low-carbon industries and technology research and development, widely promote the green procurement system, and help enterprises efficiently produce green and low-carbon products and develop technologies[2]; Research and formulate tax incentives for enterprises to reduce CEs, and encourage the development of a green and low-carbon economy; Strengthen the constraint and guidance of prices, strictly prohibit the implementation of electricity price discounts on "double high" industries to increase their production costs, and increase the implementation of policies such as differentiated electricity prices to reflect the incentive effect of prices. In terms of carbon markets, carbon trading serves as a key market regulation mechanism that provides economic incentives for emission sources (such as factories, power plants, etc.) by effectively allocating CE rights, achieving carbon reduction targets at low cost. Since the launch of the national carbon market, the power industry has been the first to be included in the national carbon trading market, with CE quotas issued to it. In the future, seven high emission industries, including petrochemicals, chemicals, building materials, steel, non-ferrous metals, papermaking, and aviation, will also gradually enter[3]. The carbon market mechanism is conducive to phasing out low-end industries, achieving industrial optimization and upgrading, and can also force enterprises to use new energy, reduce the demand for CE rights, and thus achieve the effect of energy conservation and carbon reduction. The implementation of these measures has injected stronger emission reduction momentum into enterprises, helping them move towards green development. Under the carbon reduction target, the impact of corporate CEs on corporate behavior and decision-making has received widespread attention. Existing research has delved into the impact of corporate CEs on their business activities, income distribution, and financing. Enterprises with high CEs need to invest heavily in equipment upgrades and technological innovations in order to meet increasingly strict environmental regulations and carbon reduction requirements. They also require higher carbon management and operating costs, such as cleaning costs, compliance costs, and litigation costs.

Zhang et al [4] further found that high CE enterprises have higher operating costs and face challenges such as lower sales growth rates and lower operating cash flows, emphasizing the difficulties faced by high CE enterprises in economic operations. The CEs of enterprises will also have an impact on income distribution. Kumari et al [5] found that the increase in carbon management costs caused by corporate CEs can affect managers' confidence in the future of the company, leading to more cautious financial decisions. It will also reduce the profitability of high CE enterprises, causing the market's attention to them to gradually weaken, leading to a decline in market value. At the same time, facing various challenges such as CE costs, market demand fluctuations, financing restrictions, and regulatory risks, the uncertainty of corporate cash flow and earnings has increased, resulting in a decrease in dividend payout ratios and limiting shareholder returns. The market is gradually favoring sustainable investment, and corporate CEs have had a profound impact on corporate financing. Ağbulut et al [6] indicates that investors recognize that environmental factors have a significant impact on the long-term performance and value of enterprises, and therefore require enterprises to compensate for CE risks to ensure the sustainability and long-term profitability of investments, thereby increasing the financing costs of enterprises. Among them, CE intensity, as a key indicator for evaluating carbon reduction performance, has become an important investment consideration factor. Institutional investors will screen potential investment targets based on the company's CE intensity, and are more inclined to invest in companies that perform well in the environment and have lower CEs[7]. Relatively speaking, for companies with high CE intensity, investors will demand higher returns to bear potential environmental and sustainability risks in the future, which increases the financing costs of the company. The research on CE scenario prediction is of great significance for authorities to formulate scientific and reasonable carbon reduction policies, and has attracted more and more scholars to invest in it. By systematically reviewing and analyzing existing domestic and foreign literature, it can be found that the methods for predicting corporate CEs mainly include the following: scenario analysis prediction method, grey prediction method, system dynamics model, and intelligent algorithm prediction model. Scenario analysis and prediction are effective methods for analyzing the macro environment of enterprises, which can flexibly consider various scenario factors and optimize quantitative models. However, this method is limited by the subjective judgment and experience of researchers, and the predicted results may be subjective [8].

On the basis of decomposing the influencing factors of enterprise energy consumption in the STIRPAT model, three scenarios are set to predict enterprise energy consumption. Using the GDIM method to decompose the CEs of enterprises, three scenarios are constructed based on this: baseline, low-carbon energy conservation, and technological breakthroughs [9]. Guo et al [10] uses the LMDI method to decompose the causes of corporate CEs, and then sets up 5 scenarios based on each factor, combined with the STIRPAT model for scenario prediction. Holland et al [11] constructs a CE calculation model for the operation process of enterprises, and sets up four scenarios to predict the CE situation in the next 40 years. The grey prediction method is mainly aimed at predicting and analyzing situations with limited quantity, relatively low data integrity and reliability. This method is usually more suitable for short-term forecasting of exponential growth and can effectively cope with situations where data is scarce or of low quality [12].

2. Multiple Regression Model

The CE flow of the power system is defined as: a virtual network flow that depends on the existence of the power flow and is used to characterize the CEs formed by maintaining the power flow of any branch in the power system. If it does not cause misunderstanding, it can be referred to as CE flow and carbon flow for short. As shown in Figure 1, a situation is considered: a coal-fired unit will generate a large amount of CO_2 when generating electricity. It is assumed that the CO_2 generated by the unit is not discharged into the air, but flows from the power supply side to the load side with the power system power flow. The resulting CE flow network is called the power system CE flow. Power system CE flow analysis is a new scheme to analyze low-carbon optimization of power system. It can reveal the impact of different energy structures, operation modes and dispatching strategies on CEs, and provide theoretical guidance for optimizing energy consumption and reducing environmental pollution.

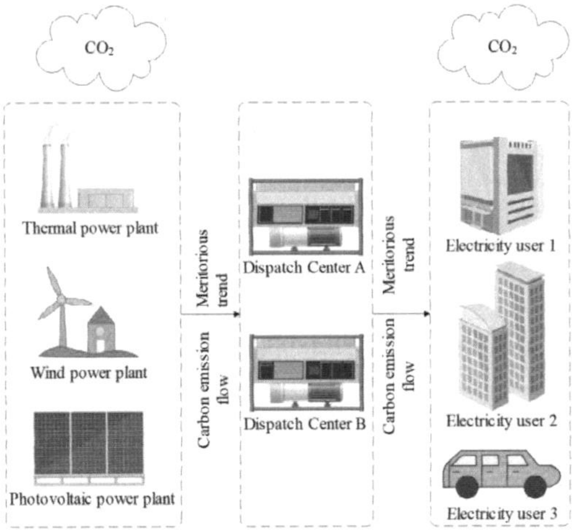

Figure 1. Schematic diagram of CE flow of power system

The relationship between power system CEs and traditional load decomposition is shown in Figure 2[13].

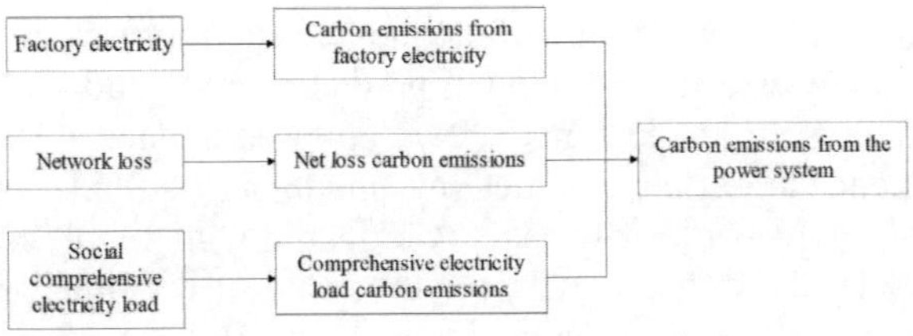

Figure 2. Relationship between CEs and thermal power supply load

It can be seen from Figure 2 that the CEs of the power system include not only the CEs during the power generation stage, but also the CEs during the power use process. In order to further explore the power CE prediction model, the power engineering CE prediction model shown in Figure 3 is proposed.

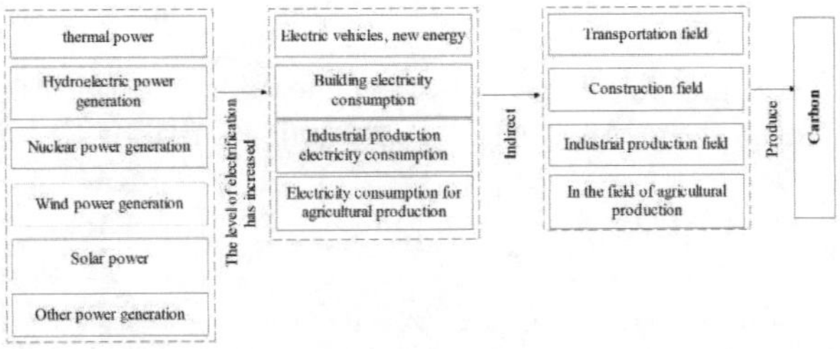

Figure 3. Schematic diagram of establishing CE prediction model for power engineering

As shown in Figure 3, power data is of great significance in the CE forecast of the power industry. By monitoring and analyzing factors such as load conditions, power generation methods and energy structure of the power system, the corresponding CEs can be calculated. Therefore, it is a feasible method to establish a regression model to explore the relationship between electricity data and CEs and realize "measuring carbon by electricity".

The CE prediction method of power engineering based on multiple regression model is shown in Figure 4, including four aspects: data collection and preprocessing, variable selection and feature engineering, model construction and parameter estimation, and model verification and prediction.

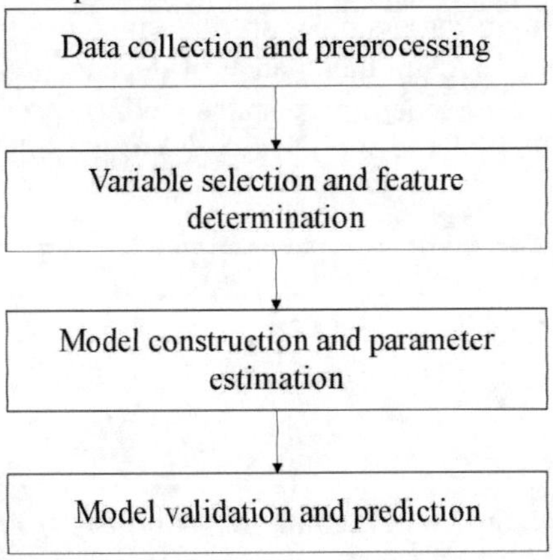

Figure 4. Flowchart of CE prediction method of power engineering based on multiple regression model

First, it is necessary to collect data on CEs and related factors of power engineering, and perform data preprocessing, including data cleaning, missing value processing and outlier value detection. The second is to select appropriate independent variables according to the influencing factors of CEs, and carry out feature engineering, including variable transformation, standardization and dimensionality reduction. The third is to establish a multiple regression model, estimate the parameters of the model by the least square method, and fit and optimize the model. The fourth is to verify the model through cross-validation and error analysis, and carry out CE prediction and trend analysis.

Multiple linear regression is a statistical method widely used in data analysis and machine learning to predict unknown dependent variable values by modeling the relationship between multiple independent variables and one dependent variable. Compared with simple linear regression, multiple linear regression can consider the influence of more independent variables on dependent variables. In this paper, only the power generation of different energy forms is selected as the independent variable, and the selected independent variables are of the same type, so a multiple linear regression model is adopted, and the formula is as follows:

$$g = a_1x_1 + a_2x_2 + a_3x_3 + a_4x_4 + a_5x_5 + a_6x_6 \quad (1)$$

In the above formula (1), g represents CEs, a_1 - a_6 is the coefficient of the multivariate linear prediction curve, and x_1 - x_6 represents thermal power generation, hydropower generation, nuclear power generation, wind power generation, solar power generation, and other power generation projects. After calling the regress function in MATLAB, the prediction curve of CEs is obtained, as shown in Figure 5.

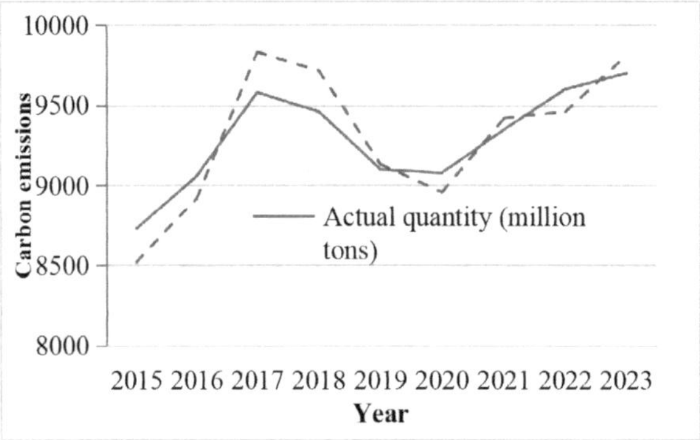

Figure 5. Actual and predicted CEs

The CE curve fitted by multiple regression is in high agreement with the real CE curve, which indicates that the model can better explain the change of dependent variable (CE) and also has good prediction ability. The prediction curve coefficients calculated by MATLAB are shown in Table 1.

Table 1. Coefficients of multivariate linear prediction curve

Coefficient	a1	a2	a3	a4	a5	a6
Numerical value	0.218	-0.087	0.899	-0.136	-1.316	74.530

The prediction curve function expression is:

$$g = 0.278x_1 - 0.087x_2 + 0.899x_3 - 0.136x_4 - 1.316x_5 + 74.53x_6 \quad (2)$$

3. Experimental Study

The daily CE data of the power industry in Liaoning Province from January 1, 2022 to September 30, 2024 in the China Carbon Accounting Database is selected. The total amount of data is 1,532, of which 80% is used as the training set and 20% as the test set. Since the LSTM neural network needs to input independent variables and dependent variables, for time series analysis, it is necessary to first transform the unsupervised learning problem into a supervised learning problem, and use the data from the previous t periods as independent variables and the data from the t+1 period as dependent variables, and use the data from the previous t periods to predict the results of the t+1 period.

LSTM, DBO-LSTM, multiple regression and traditional time series ARIMA prediction models are used to predict the CEs on the test set, respectively, and the prediction results are shown in Figure 6.

The prediction results of the ARIMA model are shown in Figure 6 (a). It can be seen from the figure that the ARIMA model has low accuracy in predicting CE trends, and its prediction effect is poor than that of the deep learning model. The main reason is that the ARIMA model can only capture linear relationships and has poor ability to capture unstable time series rules. The LSTM prediction results are shown in Figure 6 (b). It can be seen from the figure that the accuracy of the LSTM model in predicting the CE trend is low, and the true value in some areas deviates greatly from the extreme point of the predicted value. The main reason is that the LSTM network has multiple gating units and memory units, which require a large number of parameters to be optimized. It is difficult to artificially select and adjust the parameters to the optimal parameters. The prediction results of DBO-LSTM model are shown in Figure 6 (c). Compared with LSTM prediction model, the volatility of DBO-LSTM curve is less, the beginning and tail are more stable, and it is more in line with the trend of the true value curve. However, the overall prediction trend shows a serious lag. The main reason is that DBO-LSTM model has higher requirements for the quantity and quality of data, and smaller data sets or low-quality data may lead to the decline of the prediction performance of the model. The prediction results of the multiple regression model are shown in Figure 6 (d). It can be seen from the figure that the multiple regression model has a good fitting effect, and its overall performance is better than that of LSTM and DBO-LSTM models. It can accurately and appropriately describe the trend and fluctuation of data, obviously reduce the lag, and respond more quickly to the future. The main reason is that VMD can reduce the complexity and instability of time series and solve the modal aliasing problem of empirical mode decomposition method.

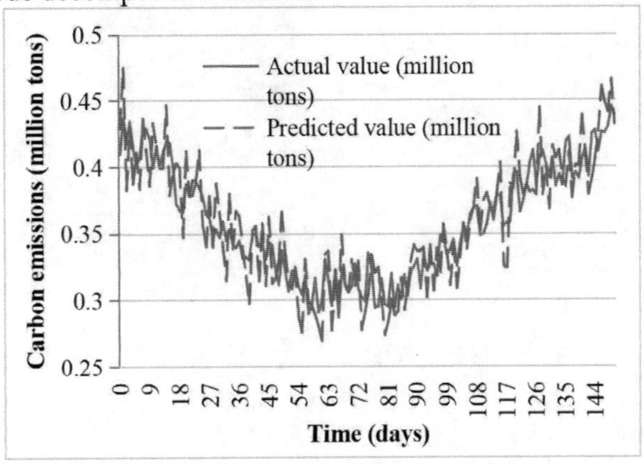

(a)

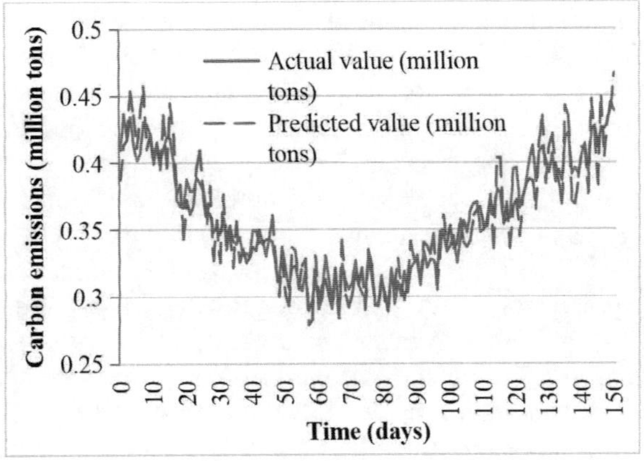

(b)

Journal of Physics: Conference Series **3036** (2025) 012006 doi:10.1088/1742-6596/3036/1/012006

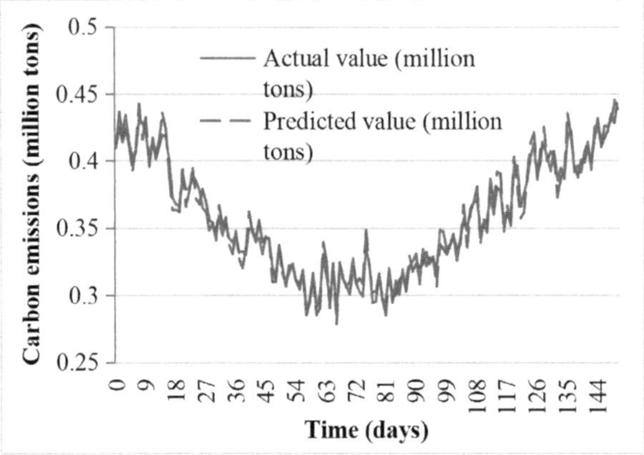

(c)

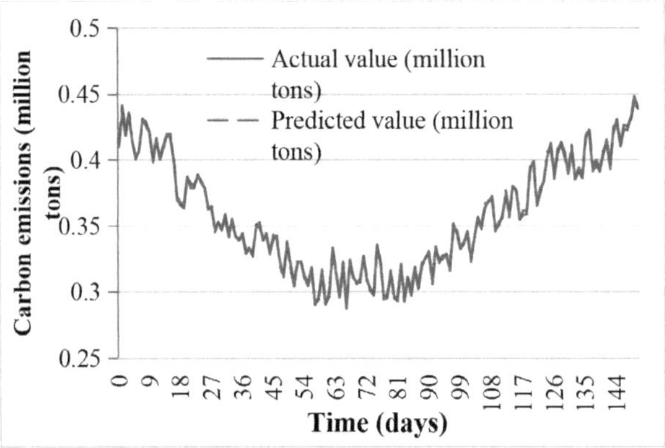

(d)

Figure 6. Prediction results of each model (a-ARIMA, b-LSTM, c-DBO-LSTM, d-VMD-DBO-LSTM)

Through experimental analysis, it can be seen that in the production-side electricity CEs, the electric energy output effect and thermal power energy structure effect show obvious promotion effect in time, while the power supply structure and thermal power energy consumption intensity show obvious inhibition effect. From the spatial point of view, the effect of electric energy production is the main reason for the spatial difference, and the effect of thermal power energy structure has the least influence on the spatial difference. In the consumption measurement of electricity CEs, population size effect and economic activity effect show obvious promotion effect in time, while electric energy consumption intensity effect and consumable electricity consumption energy intensity effect show obvious inhibition effect. From the spatial point of view, population size effect, economic activity effect and electric energy consumption intensity effect all have great influence on the spatial differences among provinces. On the whole, the consumption effect promotes the growth of CEs in the power industry, while the production effect inhibits the growth of CEs. Compared with the CEs on the consumption side, the CEs calculated from the perspective of the production side will reduce the promotion effect of the consumption effect of the eastern coastal areas with a large amount of electric energy transfer on CEs, while the consumption effect of the western areas with a large amount of electric energy transfer will reduce the promotion effect of CEs. Moreover, the promotion effect of CEs increases, and it is more reasonable to analyze the driving factors related to consumption of electricity CEs from the perspective of consumption side. Compared with CEs on the production side, CEs calculated from the perspective of the consumption side will enhance the inhibitory effect of production effects in the eastern coastal areas on CEs. Therefore, it is more reasonable to analyze the driving factors related to production of electricity CEs from the perspective of the production side. In terms of technical efficiency, the thermal power energy efficiency effect and thermal power emission efficiency effect in most provinces have inhibited the growth of production-side CEs, while the thermal power energy-saving technology effect and thermal power emission reduction technology effect have promoted the growth of production-side CEs. In addition, electricity consumption efficiency and electricity conservation technologies in most provinces have suppressed the increase of consumer-side CEs.

4. Conclusion

According to the analysis results of the model proposed in this paper, the correlation between CEs and power generation of thermal power generation, wind power generation, nuclear power generation, solar power generation and other power generation modes varies according to their characteristics. Thermal power generation mainly relies on fossil fuels, such as coal, natural gas and oil. Moreover, the greenhouse gases such as carbon dioxide and carbon monoxide produced by its combustion lead to higher CEs. The power generation of thermal power generation is related to fuel supply and market demand, and shows a clear positive correlation with power generation. In addition, the correlation between different types of energy generation and CEs depends on factors such as energy sources, production processes, and equipment technology. Therefore, in the process of promoting sustainable development and protecting the environment, priority should be given to using low-carbon, energy-efficient and energy-saving power generation methods.

References

[1] Yu, H., Yang, Y., Li, B., Liu, B., Guo, Y., Wang, Y., ... & Meng, R. 2023 *Scientific Reports*, **13**(1), 5568-5577.
[2] Luo, J., Zhuo, W., Liu, S., & Xu, B. 2024 *IEEE Access*, **12**(1), 14690-14702.
[3] Wu, Z., Wang, Z., Yang, Q., & Li, C. 2024 *Energies*, **17**(2), 472-484.
[4] Zhang, C., Ma, T., Shi, C., & Chiu, Y. H. 2023 *Energy & Environment*, **34**(6), 1910-1936.
[5] Kumari, S., & Singh, S. K. 2023 *Environmental Science and Pollution Research*, **30**(55), 116601-116616.
[6] Ağbulut, Ü. 2022 *Sustainable Production and Consumption*, **29**(2), 141-157.
[7] Bamisile, O., Obiora, S., Huang, Q., Yimen, N., Idriss, I. A., Cai, D., & Dagbasi, M. 2021 *International Journal of Hydrogen Energy*, **46**(2), 2755-2773.
[8] Gao, M., Yang, H., Xiao, Q., & Goh, M. 2022 *Renewable Energy*, **181**(1), 803-819.
[9] Wang, Y., Qiu, J., & Tao, Y. 2021 *IEEE Transactions on Power Systems*, **37**(4), 2894-2905.
[10] Guo, L. N., She, C., Kong, D. B., Yan, S. L., Xu, Y. P., Khayatnezhad, M., & Gholinia, F. 2021 *Energy Reports*, **7**(1), 5431-5445.
[11] Holland, S. P., Kotchen, M. J., Mansur, E. T., & Yates, A. J. 2022 *Proceedings of the National Academy of Sciences*, **119**(8), e2116632119- e2116632130.
[12] Ozcan, B., & Ulucak, R. 2021 *Nuclear Engineering and Technology*, **53**(6), 2056-2065.
[13] Magazzino, C., Mele, M., & Schneider, N. 2021 *Renewable Energy*, **167**(1), 99-115.

Acknowledgment

This article is supported by the Science and Technology Project (522814240007) of State Grid Qinghai Electric Power Company.

Analysis and optimal configuration of integrated energy system in parks based on artificial intelligence

Lichun Yan*, Liang Zhu, Shengxuan Liao

Longyuan (Beijing) New Energy Engineering Design&Research Institute Co., Ltd., Beijing 100034 China

Corresponding email: lisayan_1999@sina.com

Abstract: The integrated energy utilization system of industrial parks is an effective way to achieve rational resource utilization and reduce energy waste. This paper proposes a reinforcement learning scheduling method for low-carbon parks that considers multiple flexible resources. By considering the coordinated operation of power-to-hydrogen and hydrogen-blended gas units with carbon capture on the energy supply side and the comprehensive evaluation of the matching degree of elements between each link, this paper measures the degree of exploitability of project-resources and the degree of responsiveness of project-demand in the integrated energy system (IES), so as to explore the flexibility of the chain ends, the coordination of the links and the overall adaptability, find the bottlenecks in the links and seek system solutions and optimization strategies. The results show that the carbon emissions and dispatching costs of the park's integrated energy system are significantly reduced by dispatching energy storage, flexible loads and guiding EV orderly charging. The flexible capabilities covered by the constructed index system can better characterize the flexible and adjustable capabilities of the park's resources, projects, and demand ends, and characterize the chain-end acceptance, coupling, and state inclusion attributes with the project as the hub.

1. Introduction

How to quickly realize low-carbon environmental protection and coordinate the contradiction between energy consumption and green sustainable development has become the focus of attention of all countries in the world [1]. Focusing on the goal of addressing climate change, countries around the world are actively formulating low-carbon development strategies, promoting the low-carbon transformation of the energy structure.

The hybrid interval prediction method uses the neural network model to get the point prediction results, and then obtains the prediction interval through the statistical method. Reference [2] proposed a combined correlation vector machine load forecasting model using Gaussian kernel density estimation, and achieved good prediction accuracy. Reference [3] combines extreme learning machine with quantile regression, and proposes a quantile regression model with multiple quantiles for the probability prediction of wind power. In reference [4], Convolutional Neural Networks(CNN) and Gated Recurrent Unit (Gru)are combined with quantile regression to obtain the predicted value of bus load at different confidence levels. Reference [5] proposes an improved Gaussian regression algorithm for interval forecasting of power load, and the experiment proves that the prediction accuracy of this method is better than quantile regression and traditional Gaussian process regression. In reference [6], the interval prediction model is established by using Long Short-Term Memory(LSTM).

This method directly constructs the prediction interval based on various machine learning models

[7]. Reference [8] uses the algorithm to optimize the double objective problem including interval coverage and interval average width, but it is difficult to balance these two objectives in many non dominated solutions of Pareto frontier. Since the objective functions of the above methods are non differentiable, the efficient gradient descent algorithm will no longer be applicable, so the heuristic algorithm can only be used to train the model[9]. Reference [10] built an interval prediction model based on Gru, proposed an adaptive optimization method based on constructing interval, provided efficient training labels for model training, and used adaptive momentum algorithm to train the model.

The reference [11] obtained the interval prediction results based on the Gaussian mixture model, and the experiment proved that its prediction accuracy is significantly better than the single distribution model. Reference [12] combines extreme learning machine with Lower and Upper Bound Estimation(LUBE) method to build a multi input and dual output model to obtain the photovoltaic power prediction interval. Reference [13] established a short-term power load interval forecasting model based on extreme learning mechanism, and used advanced gray wolf optimization algorithm and ant lion algorithm to optimize the model parameters to obtain a higher precision forecasting interval. Reference [14] uses the nonparametric method of kernel function extreme learning machine to construct the load interval forecasting model to further improve the forecasting accuracy.

Reference [15] uses conditional value at risk to measure the uncertainty of source load, establishes a stochastic programming model, and provides an economic scheduling scheme under the expected risk level.

The flexible supply-demand relationship of different energy subsystems is converted through multi energy coupling equipment. A higher proportion of highly volatile renewable energy is used to replace the traditional controllable thermal power units, which greatly increases the demand for flexibility. The park contains a variety of energy subsystems, which are coupled through flexible resources. The improvement of the flexibility of a single energy subsystem may lead to the reduction of the flexibility of other energy subsystems. Therefore, characterizing the supply and demand of flexibility is conducive to the reliable operation of multiple energy sources within the system.

This paper focuses on the research of carbon transaction cost model, interval prediction, robust optimization method and other problems, to realize the economic, low-carbon and reliable operation of the comprehensive energy system under the carbon trading market.

2. Artificial intelligence energy optimization algorithm

The IES of the Industrial Park combines the information network, distribution network and cooling and heating pipe network to manage the distributed energy stations such as generating units, distributed photovoltaic power stations such as building roofs and cooling, heating and cooling energy users such as power load, cooling and heating load, steam load, etc. in real time, and organically connects the Supervisory Control And Data Acquisition (SCADA) system, load integrated dispatching system and user side energy consumption status visual management through the terminal mobile platform, so as to realize data sharing and interaction. Through the terminal mobile platform, the park staff can also conduct unified management and regulation of the entire IES, timely adjust the energy strategy. The platform can also provide energy consulting, energy-saving scheme design and other services for enterprises in the park, helping enterprises reduce energy costs and improve economic benefits

The main information of reinforcement learning model includes state space s, action space a, reward value R and discount factor γ. The state space and action space are related to the decision variables and some known quantities in the park integrated energy model.

The action space represents the set of changes of action variables in each time step. The action variables include: wind power PV, Hydrogen Gas Turbine (HGT) unit output, electric boiler, hydrogen blending ratio, flue gas split ratio, AC unit output, battery action, power/cooling/heat reduction load, thermal cooling load after considering thermal inertia, and EV charging and discharging capacity. T period action space set is represented as[16]:

$$A^t=\left[P_{Wvh}^t,P_V^t,P_{hgt,e}^t,P_{eb}^t,\varepsilon^t,C_{ac}^t,\delta^t,P_e^t,L_{e,id}^t,L_{H,id}^t,L_{C,id}^t,L_C^t,L_H^t,P_{EV}^t\right] \tag{1}$$

In the formula, A^t represents the action space set in period t. P_{Wvh}^t is the power provided to EL by wind and solar power in period t, $P_{hgt,e}^t$ is the power output of HGT unit in period t, P_{eb}^t is the power consumption of EB in period t. ε^t is the hydrogen blending ratio, C_{ac}^t is the cooling output of AC unit in period t, δ^t is the flue gas diversion ratio, P_e^t is the state of the charging and discharging action of the storage battery, $L_{e,id}^t$ is the power interruption load, $L_{H,id}^t$ is the heat interruption load, $L_{C,id}^t$ is the cooling load, L_C^t is the cooling load of the system in period t, L_H^t is the heating load of the system in period t, and P_{EV}^t is the electric power consumed by the EV device in period t.

State space generally considers the factors that will affect decision-making. In this paper, the state space includes the electricity load, heating load, cooling load, wind and solar forecast value, electricity price, gas price, purchased electricity in the next period and the action amount in the previous period, namely:

$$S^t=\left[L_{e,DR}^t,P_H^t,P_C^t,f_W^t,f_v^t,c_{buy,e}^t,c_{buy,g}^t,P_{grid}^t,A^t\right] \tag{2}$$

S^t represents the state space set of period t, $L_{e,DR}^t$ represents the electric load after price-type demand response in period t, P_H^t represents the system heat load demand in period t, P_C^t represents the system cooling load demand in period t, and f_W^t and f_v^t represent the wind and light forecast values in period t, respectively. $c_{buy,e}^t,c_{buy,g}^t$ is respectively the electricity and gas purchase price in period t, P_{grid}^t is the electricity purchase amount of the low-carbon park in period t, and A^t is the action space set in period t.

Reinforcement learning continuously learns and updates its own strategies through the agent, so as to maximize the expected reward value. The reward function rt is shown in formula (3):

$$r^t=-\partial F^t+k \tag{3}$$

In the formula, ∂ is the coefficient, F^t is the total cost in period t, and k is the correction parameter that makes the reward function tend to a positive value.

3. Model construction and test

3.1 Model construction

The production mode and consumption structure of electric energy have undergone significant changes. The electricity-heat-cold IES can realize the mutual conversion of electricity, heat and cold energy.It is necessary to deeply analyze its basic structure and construct its internal unit models.

Journal of Physics: Conference Series **3036** (2025) 012007 doi:10.1088/1742-6596/3036/1/012007

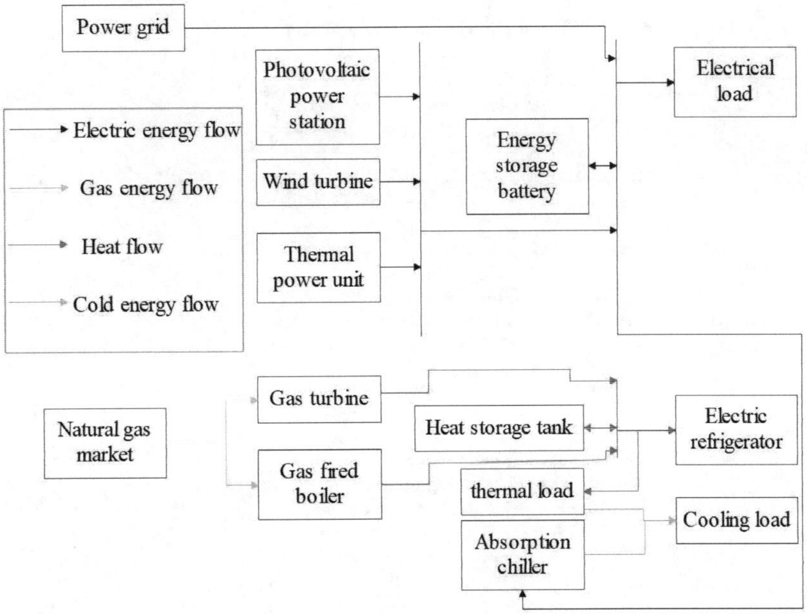

Figure 1. Framework diagram of IES considering electricity-heating-cooling

Figure 1 above is a basic framework diagram of an IES including electricity, heating and cooling. The electrical load is coordinated and provided by gas turbines, thermal power generating units, energy storage batteries, and power grids. The heat load is coordinated by gas boiler. The energy interaction of the multi-microgrid group is shown in Figure 2.

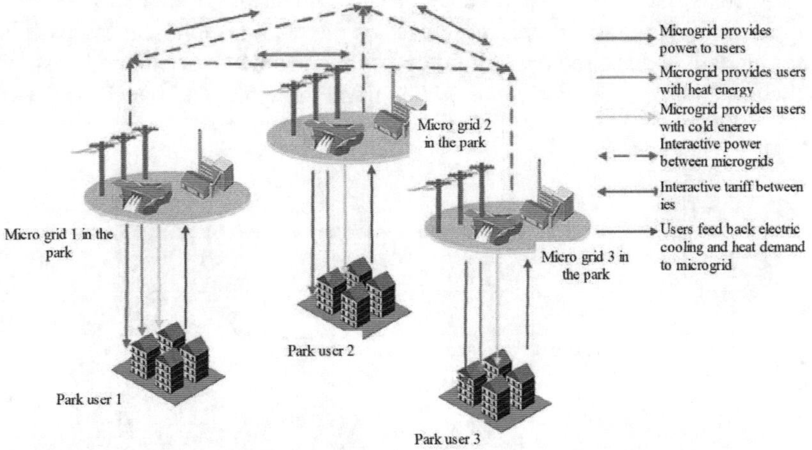

Figure 2. Schematic diagram of energy interaction of multi-microgrid group

The direct and indirect energy consumption of production and living behaviors is analyzed through energy utilization and conversion links. Taking the direct energy demand of the park as the link, splitting the energy conversion and utilization links in the park's consumption process can more accurately plan energy conversion projects for the park's real energy demand, and can comprehensively evaluate the emission reduction benefits and costs of the projects. The overall analysis process is shown in Figure 3.

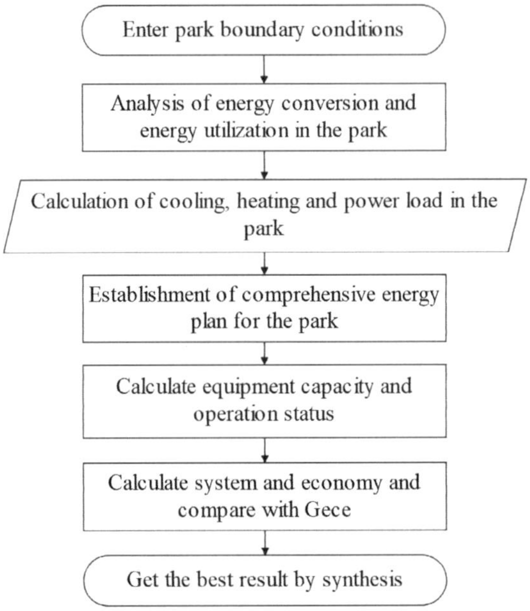

Figure 3. Analysis flow of comprehensive energy scheme in the park

3.2 Example design

Taking the low-carbon park comprehensive energy system shown in Figure 1 as an example, the effects of various flexible resources on carbon reduction and cost reduction are analyzed. The gas purchase price at the gas station is 0.512 yuan/kWh. The number of EV in the low-carbon park is set to 500, and the upper limit of the total charging power of the charging piles in the park is 200 kW. The disordered charging daily load of the park simulated by the Monte Carlo method is shown in Figure 4, and the typical wind/light/load forecast curve is shown in Figure 5.

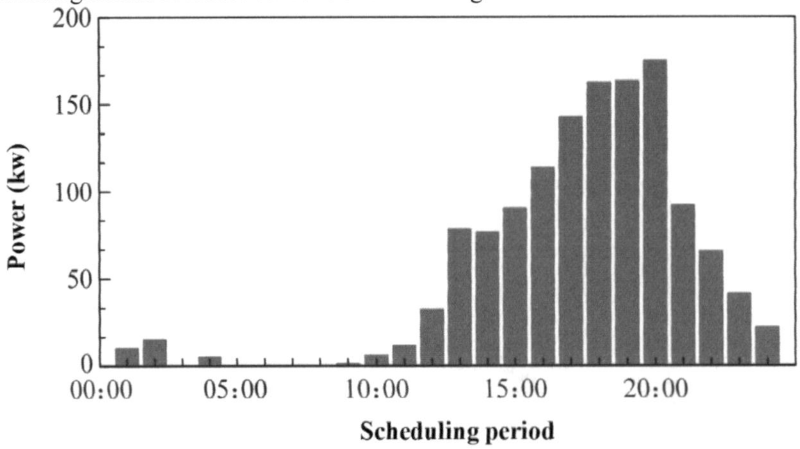

Figure 4. Daily load of disorderly charging of electric vehicles

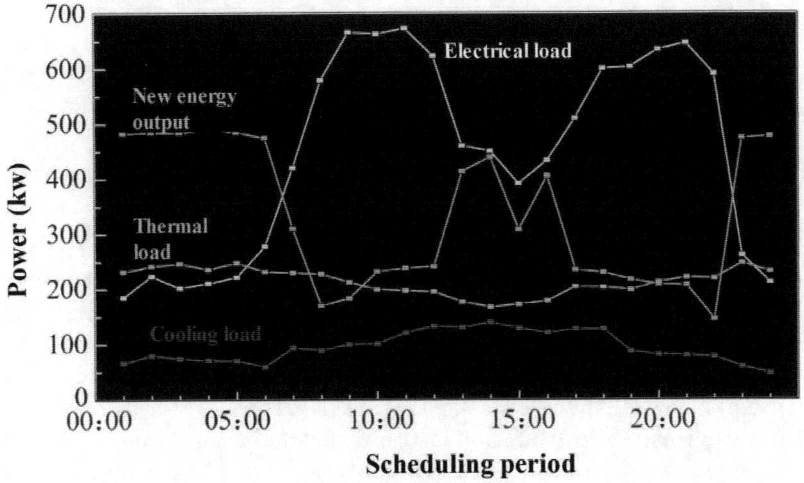

Figure 5. Wind/light/load forecast curve

In the algorithm, the reward discount factor is 0.9, the number of neurons is 256, the noise is 0.2, and the delay policy update frequency is 2. It can be seen from Figure 6 that the reinforcement learning agent in this paper converges after 1600 iterations, and the final reward value is stable at around 52. The reward value of the agent increases significantly in 500 ~ 600 iterations, and decreases significantly in 600 ~ 700 iterations. The agent's optimization range is large, and then the double Q network's restricted optimization space becomes more conservative, resulting in a decrease in reward value. In the 700-1600 iterations, the agent's reward value gradually and slowly reaches stability. After stabilization, the agent's oscillation amplitude is significantly reduced.

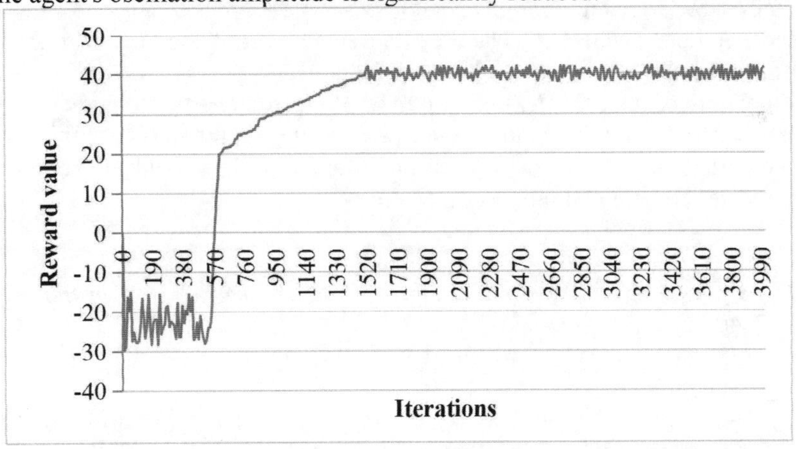

Figure 6. Reward value of reinforcement learning algorithm training process

This paper designs the following scenarios for model performance analysis. Scenario 1: The system flexible resources are not considered for scheduling, that is, EV orderly charging and comprehensive demand response for heating, cooling and electricity are not included; Scenario 2: Only a single flexible resource is considered, that is, comprehensive demand response is considered but EV orderly charging is not considered; Scenario 3: Only a single flexible resource is considered, EV orderly charging is considered but comprehensive demand response is not considered; Scenario 4: Multiple types of demand response resources are considered. The above scenarios all use the TD3 algorithm based on behavior cloning for offline training, use the same historical data set for training, and then use the scheduling day data in Figure 5 to test the trained model. The scheduling results of each scenario are shown in Table 1.

Table 1. Comparison of optimized scheduling results

Scene	Cost/yuan	Carbon emissions/kg	Power purchase cost/yuan	Gas purchase cost/yuan
Scenario 1	10,814.87	6,849.41	5,126.22	3,933.52
Scenario 2	9,887.32	6,026.09	4,132.64	3,688.24
Scenario 3	10,103.89	5,888.73	4,321.86	4,015.03
Scenario 4	8,809.68	5,610.46	2,979.62	3,806.32

Compared with Scenario 1. The reason is that the comprehensive demand response in scenario 2 reduces the peak-to-valley difference in cooling, heating and electricity loads, thereby reducing the system's gas and electricity purchase costs. In scenario 3, due to the implementation of orderly charging of EVs, centralized charging during the peak period of system power consumption and peak period of electricity prices is avoided, and discharge is carried out during peak periods, thereby reducing the system's power purchase demand. For scenario 4, the dual measures of orderly EV charging and comprehensive demand response are adopted. This scenario greatly reduces the peak-to-valley difference of the system through multi-type flexible resource regulation. In addition to reducing the system scheduling cost, the system has excess electricity to supply carbon capture devices, which increases the system's carbon capture capacity and greatly reduces the system's carbon emissions.

4. Conclusion

A reinforcement learning environment is established through this model, and an improved algorithm is used to optimize it. This paper draws the following conclusions: The carbon emissions and dispatching costs of the park's IES can be effectively reduced by dispatching energy storage, flexible loads, and guiding EV orderly charging. Secondly, the flexible capabilities covered by the constructed index system can better characterize the flexible and adjustable capabilities of the park's resources, projects, and demand ends, and characterize the chain-end acceptance, coupling, and state inclusion attributes with the project as the hub.

Since this study only considered flexible resources of electric energy storage on the energy storage side and did not consider carbon capture devices containing liquid storage, further research can be carried out by comprehensively considering flexible resources of heat, cold, and air energy storage and carbon capture devices containing liquid storage.

References

[1] Li, Y., Wang, B., Yang, Z., Li, J., & Li, G. 2021 *IEEE Transactions on Industry Applications*, **58**(2), 2640-2651.
[2] Zhang, Y., Xie, X., Fu, W., Chen, X., Hu, S., Zhang, L.,et al. 2022 *IEEE Transactions on Circuits and Systems II: Express Briefs*, **70**(1), 246-250.
[3] Cao, M., Shao, C., Hu, B., Xie, K., Li, W., Peng, L., & Zhang, W. 2021 *IEEE transactions on sustainable energy*, **13**(1), 290-301.
[4] Liu, L. N., & Yang, G. H. 2022 *IEEE Transactions on Industrial Informatics*, **18**(10), 6569-6580.
[5] Chen, C., Li, Y., Qiu, W., Liu, C., Zhang, Q., Li, Z., ... & Yang, L. 2022 *IEEE Transactions on Sustainable Energy*, **13**(4), 1994-2011.
[6] Lin, S., Liu, C., Shen, Y., Li, F., Li, D., & Fu, Y. 2021 *IEEE Transactions on Smart Grid*, **13(**1), 202-212.
[7] Meng, Q., Zu, G., Ge, L., Li, S., Xu, L., Wang, R., et al. 2022 *Applied Sciences*, **12**(23), 12309-12320.
[8] Peng, Q., Wang, X., Kuang, Y., Wang, Y., Zhao, H., Wang, Z., & Lyu, J. 2021 *CSEE Journal of Power and Energy Systems*, **7**(5), 911-921.
[9] Cao, J., Yang, Y., Qu, N., et al. 2024 *Scientific reports*, **14**(1), 4129-4140.
[10] Wang, L., Tan, M., Chen, J., & Liao, C. 2023 *Applied Intelligence*, **53**(9), 10273-10289.

[11] Zhou, X., Ma, Y., Wang, H., Li, Y., Yu, J., & Yang, J. 2022 *Energy Reports*, **8**(1), 527-535.
[12] Liu, X. 2022 *International Journal of Energy Research*, **46**(14), 20535-20555.
[13] Su, J., Chiang, H. D., Zeng, Y., & Zhou, N. 2021 *IEEE Transactions on Smart Grid*, **12**(4), 3004-3015.
[14] Zhou, Y., Wei, Z., Shahidehpour, M., & Chen, S. 2021 *IEEE Transactions on Power Systems*, **36**(4), 3574-3584.
[15] Radchenko, A., Scurtu, I. C., Radchenko, M., Radchenko, R., Forduy, S., & Zubarev, A. 2022 *Thermal Science*, **26**(1 Part A), 185-194.
[16] Fan, W., Liu, Q., & Wang, M. 2021 *Energies*, **14**(16), 4740-4755.

Acknowledgments
Longyuan Electric Power Technology Project (LYH-2022-01).

Journal of Physics: Conference Series **3036** (2025) 012008 doi:10.1088/1742-6596/3036/1/012008

An intelligent carbon emission measurement model based on high-frequency power data

Rongfu Zhang [1]*, Guangde Zhang[1], Liangyu Han[2], Shengbo Li[1], Chen Fang [1]

[1]Information and Communication Company of State Grid Qinghai Electric Power, Xining, Qinghai 810000, China

[2]State Grid Qinghai Electric Power Company, Xining City of Qinghai, 810000, China

Corresponding email: santiagoPangZL@163.com

Abstract. This paper introduces genetic learning strategy into the traditional artificial bee colony algorithm, proposes an evolutionary artificial bee colony algorithm, and gives full play to its global search ability to automatically search for the optimal parameters of the random forest regression model. At the same time, this paper proposes a random forest regression prediction model based on the optimization of the evolutionary artificial bee colony algorithm, so as to accurately predict the future development trend of carbon emissions in the power industry and provide strong support for the formulation of energy conservation and emission reduction policies. The results of the practice show that the real-time perception system of urban carbon emissions based on power big data can significantly improve the accuracy and timeliness of carbon emission data, and can provide strong support for the competent authorities to formulate carbon emission reduction policies. In the future, we can continue to deepen research, promote technological innovation and application, contribute more wisdom and strength to urban carbon emission management, and promote cities towards a low-carbon, green and sustainable future.

1. Introduction

International experience and academic research have shown that strong and predictable carbon prices play an important role in effectively managing the process of carbon neutrality and transformation. The price adjustment methods of the carbon market cover a variety of ways, including carbon emissions trading mechanism, carbon tax and carbon credit. Among them, the carbon emissions trading mechanism has become the core policy means to promote carbon emission reduction.

At present, there are 28 carbon emission trading markets operating in the world, accounting for about 17% of the total global emissions. Article 6 of the Paris agreement stipulates that countries can achieve their emission reduction targets through the market mechanism. Article 6.2 stipulates that the emission reduction achievements between countries can be transferred internationally; Article 6.4 is intended to establish an international carbon emission trading market supervised by United Nations agencies [1]. The first performance period of Washington's total carbon emission control and trading system will start in January 2023, which will cover about 70% of greenhouse gas emissions. The energy crisis has not affected the stable operation of the EU carbon market, which once rose to 95 euros/ton, and will launch a parallel carbon market in 2027, covering road transportation, construction and other fields. In addition, 13 countries or regions in Latin America and Asia are planning to implement or consider implementing the carbon emission trading market mechanism [2].

Journal of Physics: Conference Series **3036** (2025) 012008 doi:10.1088/1742-6596/3036/1/012008

As the key factor of power production end, the composition of power production structure is directly affected by the generation cost of different types of energy [3]. At present, the type of energy consumed by the power generation side plays a decisive role in carbon emissions in the process of power generation. In order to achieve the national "double carbon" goal, it is particularly critical to accurately assess the carbon footprint of power consumption and promote the transformation of power consumption sector to low carbon. Under the guidance of the goals of "carbon peaking" and "carbon neutralization", the future power structure will pay more attention to the application of clean energy, with wind power, solar power and other non fossil energy as the main body. With the construction of the new power system, the proportion of non fossil energy in the power generation end will increase significantly, and the power industry will gradually move towards deep low-carbon or even zero carbon [4]. At the same time, with the increasing demand of power consumption sectors such as industry, transportation and construction, the electrification level of the whole society will continue to improve. To meet this demand, we will rely more on low-carbon electricity to achieve the goal of reducing carbon and decarbonizing in the future [5].

There are still the following problems in the application of carbon emission factors: first, the data is not updated in time, second, the spatial-temporal resolution is not reflected enough, and third, the green environmental value of clean electricity has not been reflected. Huang et al [6] has verified that the carbon emissions of the thermal power industry in different regions and provinces have great differences through geographically weighted regression model, and there are positive spatial correlation and spatial dependence. This conclusion also shows that the direct use of the grid carbon emission factor can not reflect the real regional carbon emission level. Chang et al [7] proposed the concept of carbon emission coefficient, and obtained the key parameters through the econometric model, using the parameter results to obtain more accurate and scientific carbon emissions of the power industry chain. When calculating the carbon emission factor of thermal power plants, Le et al [8] proposed to allocate the carbon dioxide emissions of self owned power plants by using the heat method, so as to accurately evaluate the carbon emission efficiency of cogeneration units. When analyzing the inter provincial carbon emission transfer, Liu et al [9] proposed a carbon emission oriented source grid load co assimilation model for China's provincial power energy system, and found that power transmission is an important factor affecting regional carbon emission through analysis, and at present, carbon emissions are transferred between provinces through long-distance EHV transmission. It is necessary to comprehensively consider the difficulty and timeliness of data collection when calculating the carbon emission factor of power grid.

Driven by the low-carbon economy, the power system is facing great pressure, but it also contains the potential to reduce CO_2 emissions. As the main industry of fossil energy consumption and CO_2 emissions in China, it is very important to study the carbon emissions in power system for low-carbon emission reduction. In recent years, the power industry has not only formulated development strategies to reduce carbon emissions, but also implemented comprehensive emission reduction plans. The implementation measures include strengthening the loss management of transmission network, promoting the construction of West to East power transmission, optimizing the operation and dispatching of power system, etc. These measures have effectively reduced carbon emissions. However, the carbon emission statistics of power system are complex and there are many factors and environmental factors. If all the data are processed simply and roughly, it is difficult to conduct a comprehensive and accurate quantitative analysis. Moreover, the carbon emission data of power system itself has the characteristics of fuzziness and complexity, and the traditional quantitative analysis method is often difficult to obtain accurate results. In this context, it is very important to establish the carbon emission index system of power system and the preliminary qualitative analysis of comprehensive evaluation model to predict carbon emissions.

In order to quantify the degree of completion of carbon emissions, how to conduct a comprehensive assessment of carbon emissions and build a carbon emissions assessment program has become the focus of research. In the research of carbon emission assessment model, different carbon emission assessment models are established. Adebayo et al [10] analyzes the technical characteristics of China's new energy power system, and constructs the basic model and technical system of the new energy power system based on "carbon emissions", aiming at the problems encountered in its development process, such as "coupling uncertainty between the source and load", "carbon emission level", "Multi-Agent Coordination", which provides a theoretical basis and practical basis for the construction and development of China's new energy power system under the background of advocating low-carbon emission reduction. Singh et al [11] based on the low-carbon economy, discusses the low-carbon benefits brought by the technology development of the power system from a new perspective, and constructs an evaluation system and model, which fundamentally clarifies the meaning and essence of the low-carbon benefits of the power system. On this basis, it is also important to select the carbon emission index of power system. Bieroza et al [12] By using the life cycle assessment method and setting a unified evaluation standard, the CO_2 emissions per unit of nuclear power generation and other power generation technologies are compared and analyzed, and the impact on the environment is comprehensively evaluated under different power generation modes. Burkhardt et al [13] explores the characteristics of the carbon emission process of the power industry, finds out several interactive factors that affect the carbon emission distribution system, and combines subjective and objective methods to construct the allocation method of carbon emission responsibility, aiming to achieve a more reasonable allocation of emission responsibility among regions, while providing theoretical support for the carbon emission decentralization of each region of the power industry. Due to the large amount of CO_2 generated in the operation of power system, some researches have begun to establish the carbon emission model and evaluation system of transmission network planning. Kartal et al [14] describes the comprehensive evaluation elements and process of transmission network planning schemes from the perspective of low-carbon energy conservation, and predicts the evaluation of transmission network planning schemes facing a high proportion of clean energy integration in the future from the aspects of subject, index, method and application of evaluation results.

2. Model construction

2.1 Regional carbon emission calculation

The research of this paper adopts the inventory compilation method to measure the carbon emissions of Qinghai Province from three dimensions: region, key industry and household. In order to use the inventory compilation method together, when calculating carbon emissions in each dimension, it is necessary to determine the corresponding emission parameters and calculate the activity level corresponding to the corresponding dimension.

For the calculation of regional carbon emissions, considering the availability of data, the carbon emissions in energy utilization are mainly measured.

Step 1: The annual value E_l of total energy consumption of each city and each industry is estimated. E represents the annual value of total energy consumption of the province and each industry, G represents the annual value of added value of the province and each industry, and G_l represents the annual value of gross domestic product of each city and each industry. Then, there is:

$$E_l = (E / G) \times G_l \quad (1)$$

Step 2: The monthly long-term equilibrium relationship model of total energy consumption and electricity consumption in various regions and industries is estimated and screened. Taking Qinghai Province as an example, Ec represents the electricity consumption of the province and various industries. After the unit root and cointegration tests of the annual series of Ec and E, the following monthly long-term equilibrium relationship model is obtained:

$$\widehat{E}_t = (\alpha_0 / 12) + \alpha_0 \times Ec_t \quad (2)$$

Step 3: The adjustment factor F is calculated. The total monthly energy consumption data for 12

Journal of Physics: Conference Series **3036** (2025) 012008 doi:10.1088/1742-6596/3036/1/012008

months of each year are calculated and recorded as $\widehat{E}_1$, $\widehat{E}_2$, …, $\widehat{E}_{12}$. Then, the adjustment factor F is:

$$F = \frac{E}{\sum_{t=1}^{12} \widehat{E}_t} \tag{3}$$

Step 4: The monthly carbon emissions are calculated. Using the adjustment factor F and the carbon emission coefficient h, the monthly carbon emissions C_t are obtained:

$$C_t = \widehat{E}_t \times F \times h \tag{4}$$

Similarly, the carbon emissions of various cities and five major industries in various cities can be calculated.

2.2 Carbon emission measurement model and prediction model

The proposed "electricity-carbon" carbon emission measurement workflow of a city is shown in Figure 1.

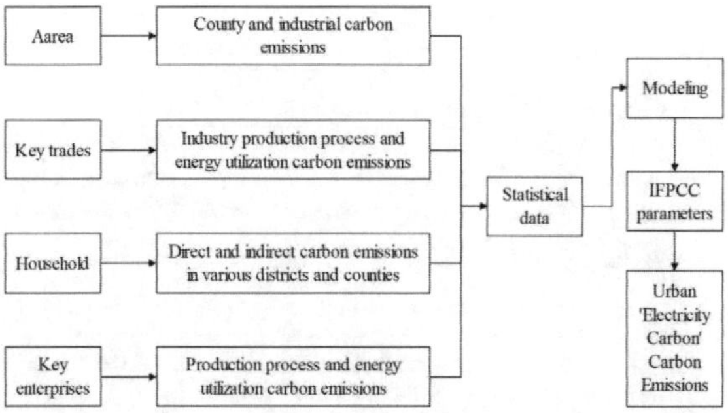

Figure 1. Work flow chart of "electricity-carbon" carbon emission calculation

Based on the power big data, a real-time carbon emission perception system for a city is built. On the basis of ensuring the security of corporate power data, the correlation between corporate power and carbon emission data is explored and verified by industry, and a case library of corporate power carbon emission correlation models is formed, thus building a bridge between power and carbon emissions and realizing the real-time conversion of power data to carbon emission data. The construction idea of a city's real-time carbon emission sensing system is shown in Figure 2, which is divided into four parts: data cleaning, model construction, "electricity-carbon" calculation, and result output.

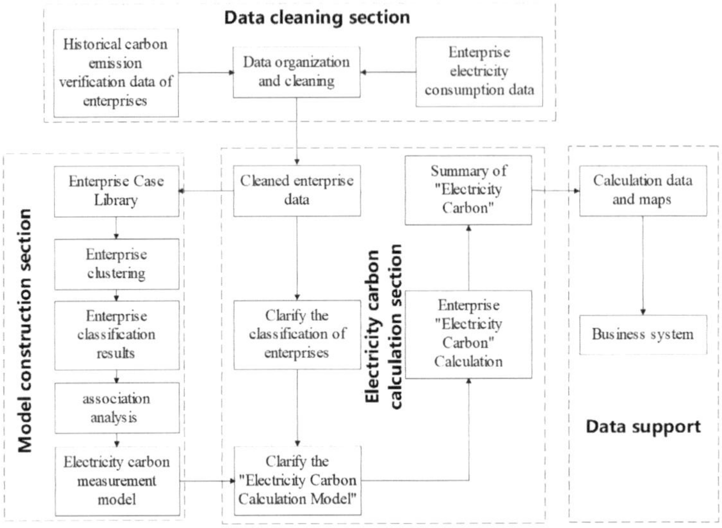

Figure 2. Construction ideas of a real-time carbon emission sensing system in a city

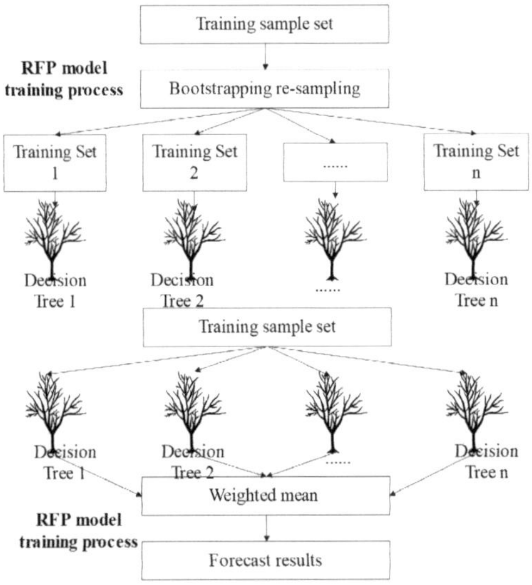

Figure 3. RFR model prediction principle

As a representative of ensemble learning theory, random forest regression (RFR) model uses the principle of "democratic voting" to set multiple decision trees to construct an enhanced predictor to solve the problem of low accuracy of a single predictor, which has been widely used in data regression prediction. When using the RFR model for data regression prediction, the original sample set is sampled by the self-help method, and several groups of training samples are randomly generated, and the decision tree is constructed based on each group of new samples. When defining the attributes of each node in the decision tree, starting from the root node of the decision tree, a number of attributes are randomly selected from all attributes of the training sample as the node attribute set, and the optimal attributes are selected according to the Gini index minimization evaluation rule, and then the binary recursive strategy is used to split the attributes and construct the nodes The prediction principle of RFR model is shown in Figure 3.

Inspired by the behavior of bee colony searching for high-quality honey source, the artificial bee colony (ABC) algorithm solves the nonlinear multi-objective and multi constraint problem by comparing the quality of honey source between artificial bee individuals. Compared with other types

Journal of Physics: Conference Series **3036** (2025) 012008 doi:10.1088/1742-6596/3036/1/012008

of group heuristic intelligent optimization algorithms, ABC algorithm has the advantages of simple structure, fast exploration speed, strong robustness and less associated parameters. However, due to the same search strategy in the employment bee honey seeking stage and the reconnaissance bee honey seeking stage, the algorithm is easy to fall into the local optimal dilemma. In order to solve this problem, a novel evolutionary artificial bee colony (EABC) algorithm is proposed by using the genetic evolution process of biological species for reference and introducing the genetic learning strategy based on the traditional ABC algorithm. Its process is shown in Figure 4.

In the honey-seeking stage of scout bees, crossover, mutation and selection operations are performed on each nectar source that meets the number of trials. Each nectar source $X_i = \left[x_{i,1}, x_{i,2}, \ldots, x_{i,D}\right]$ represents a possible solution to the optimization problem, where $i \in \{1,2,\ldots,S\}$ represents the number of artificial bees, S is the maximum number of artificial bees, and D represents the dimension of the optimization problem. EABC algorithm can effectively improve local search ability and convergence speed on the premise of retaining the good global search ability of traditional ABC algorithm.

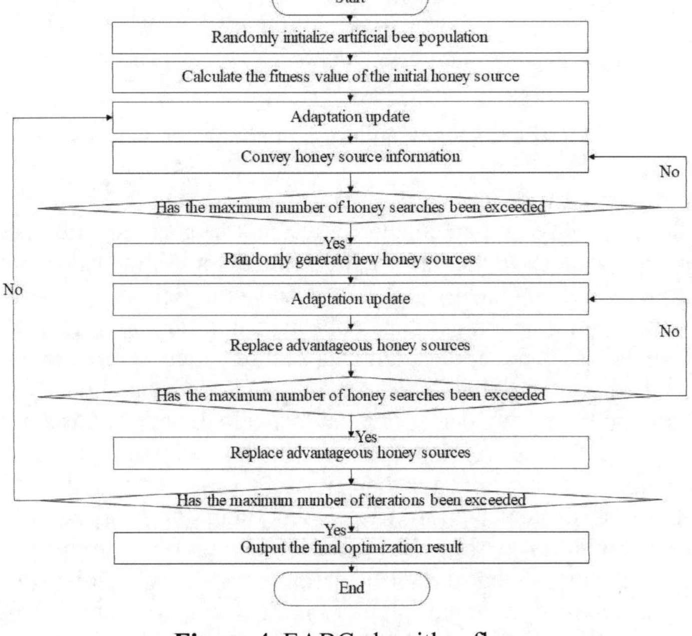

Figure 4. EABC algorithm flow

In order to ensure the stability and reliability of the carbon emission prediction results in the power industry, the traditional ABC algorithm is improved, and the EABC algorithm is proposed. The entire prediction process is shown in Figure 5. The excellent global optimization ability of the algorithm is fully utilized, and the key influencing parameters in the RFR model are automatically searched, and an RFR prediction model based on EABC algorithm optimization is proposed, thereby overcoming the disadvantages caused by artificial subjective setting of model parameters.

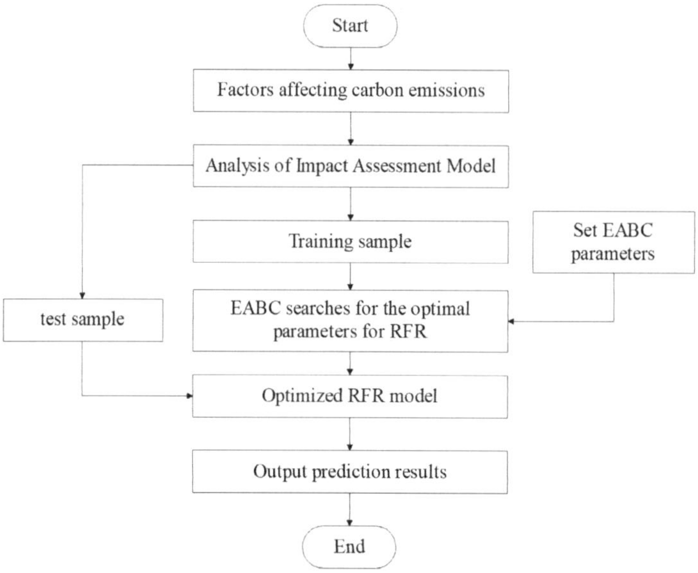

Figure 5. Carbon emission prediction process

3. Experimental Test

In order to verify the feasibility and calculation effect of the method, this study conducts a case analysis on a city in a hot-summer and cold-winter area, and applies the calculation method based on the city's multi-year public building energy verification and audit reports.

In the application stage information and data collection link, the basic information and electricity consumption data of buildings in the application data set are sorted out separately, and the electricity consumption data is sorted out according to two scales: annual data and monthly data. The sorted data is used as model input, the bivariate calculation model is called, and the carbon emission calculation value is calculated by using the electrification rate and comprehensive factor calculation value. After the calculation is completed, the calculated carbon emission values of all buildings are compared with the actual values to verify the calculation effect. According to statistics, there are some public buildings (mostly office buildings) with an electrification rate of 100% in this area. In order to avoid such buildings affecting the evaluation of the calculation effect, only buildings with an electrification rate of less than 100% are considered in the comparison. The comparison results are shown in Figure 7. Due to the small scale of the monthly model and the large sample size, only the sample comparison results numbered 1 to 200 are displayed. It can be seen from the figure that the measured values of carbon emissions are generally similar to the actual values.

Journal of Physics: Conference Series 3036 (2025) 012008 doi:10.1088/1742-6596/3036/1/012008

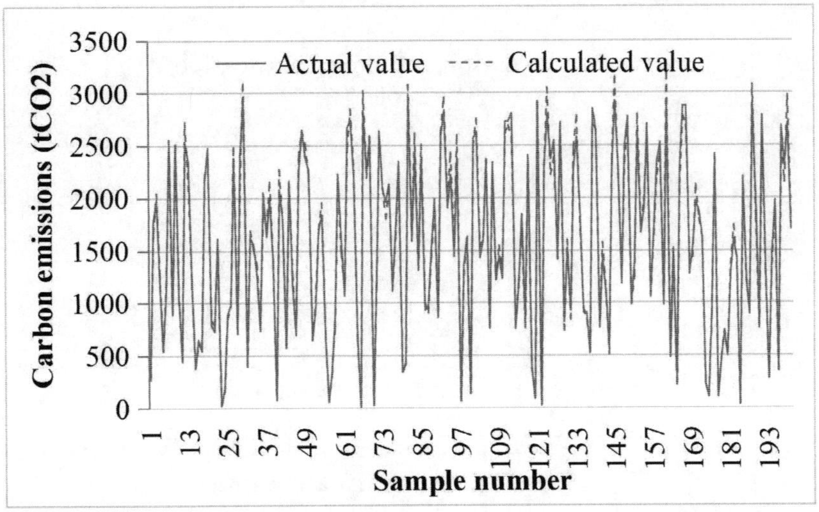

(a)

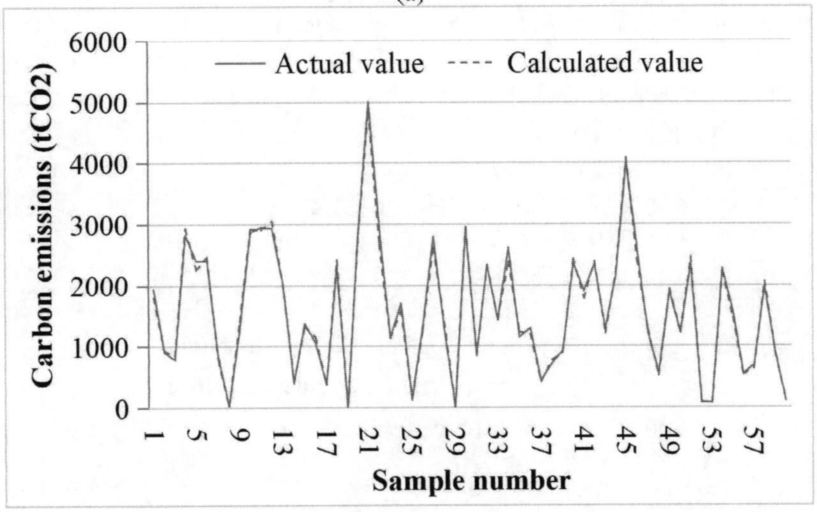

(b)

Figure 6. Comparison results of measured samples(a-monthly carbon emission; b-annual carbon emission)

When EABC algorithm is used to automatically optimize the parameters of RFR model, the nectar fitness value gradually converges with the iterative process of bee colony, as shown in Figure 7. After the search, the optimal number of decision trees of RFR model determined is 137, and the optimal number of splitting variables is 8. After setting the relevant parameters of the model, the model with optimized parameters is used to predict the carbon emissions from 2019 to 2024, and the final result is shown in Figure 8.

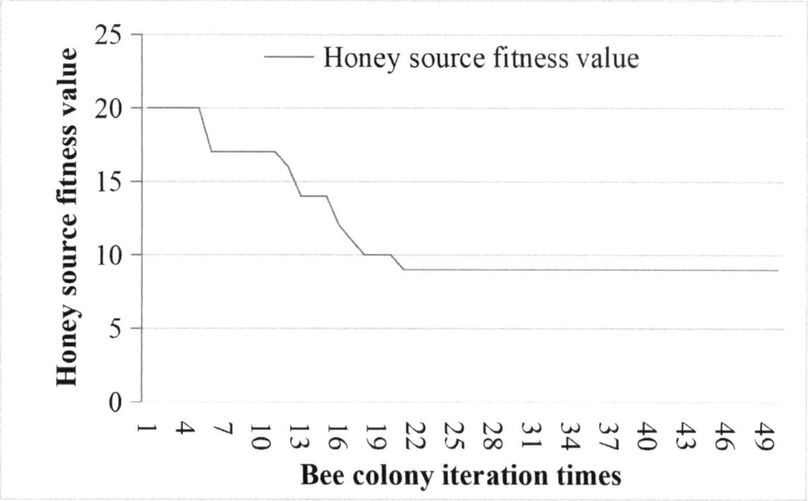

Figure 7. Iterative convergence curve of algorithm

From the prediction results, the error between the predicted carbon emissions and the actual carbon emissions is very small, and the trends of the two curves are basically consistent, which verifies the stability and reliability of the prediction model

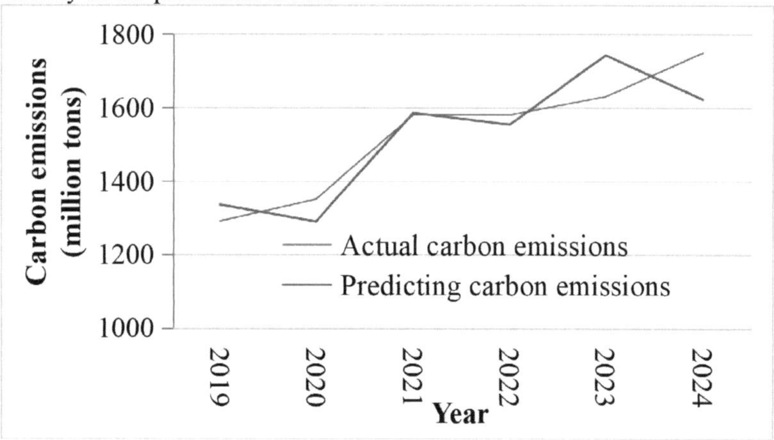

Figure 8. Model prediction results

4. Conclusion

On the basis of traditional artificial bee colony algorithm, genetic learning strategy is introduced, and evolutionary artificial bee colony algorithm is used to automatically optimize the number of decision trees and the number of split variables in random forest regression model, and a prediction model is proposed. The verification results show that the model can accurately predict the development trend of carbon emissions in the power industry and has good stability and reliability. Moreover, compared with other widely used prediction models, it has obvious advantages and can provide strong support for the formulation of energy conservation and emission reduction policies in the power industry across the country and in various provinces and cities, and provide a certain reference for solving similar data regression prediction problems.

References

[1] Liu, H., & Zhai, J. 2025 *Processes*, **13**(2), 595-606.
[2] Saadaoui, H., Dogan, M., & Omri, E. 2024 *Environmental Economics and Policy Studies*, **26**(2), 239-261.

[3] Kartal, M. T., Pata, U. K., Kılıç Depren, S., & Erdogan, S. 2024 *International Journal of Sustainable Development & World Ecology,* **31**(1), 89-107.
[4] Lu, Q., Duan, H., Shi, H., Peng, B., Liu, Y., Wu, T., ... & Wang, S. 2022 *npj Urban Sustainability*, **2**(1), 34-45.
[5] Kartal, M. T., Ali, U., & Nurgazina, Z. 2022 *Environmental Science and Pollution Research*, **29**(55), 83824-83838.
[6] Huang, Y., Wu, H., & Zhu, H. 2021 *Environmental Science and Pollution Research*, **28**(37), 51908-51927.
[7] Chang, T., Hsu, C. M., Chen, S. T., Wang, M. C., & Wu, C. F. 2023 *Economic Analysis and Policy*, **79**(1), 319-342.
[8] Le, T. H. 2022 *Renewable Energy*, **195**(1), 442-454.
[9] Liu, G., Liu, J., Zhao, J., Qiu, J., Mao, Y., Wu, Z., & Wen, F. 2022 *IEEE Transactions on Industrial Informatics*, **19**(2), 1401-1412.
[10] Adebayo, T. S., & Ullah, S. 2024 *Sustainable Development*, **32**(1), 244-259.
[11] Singh, R., Paniyil, P., & Zhang, Z. 2022 *IEEE Power Electronics Magazine*, **9**(2), 39-47.
[12] Bieroza, M., Acharya, S., Benisch, J., Ter Borg, R. N., Hallberg, L., Negri, C., ... & Kirchner, J. W. 2023 *Environmental Science & Technology*, **57**(12), 4701-4719.
[13] Burkhardt, J., Gillingham, K. T., & Kopalle, P. K. 2023 *Management Science*, **69**(12), 7784-7798.
[14] Kartal, M. T. 2023 *Environmental Science and Pollution Research*, **30**(3), 6864-6874.

Acknowledgments
This article is supported by the Science and Technology Project *Research on the Technical Standards for "Electricity Carbon Model" Calculation Based on High Frequency Power Data* (522814240007) of State Grid Qinghai Electric Power Company.

Journal of Physics: Conference Series **3036** (2025) 012009 doi:10.1088/1742-6596/3036/1/012009

Early Sign Recognition and Intelligent Warning Technology of Arc Fault in Electric Bicycle Charging Facilities

Pan Pan

University of Science and Technology of China

Corresponding author's e-mail:14156993@qq.com

Abstract. With the increase in the use of electric bicycles, the safety hazards faced by their charging facilities are becoming increasingly serious, especially arc faults. Arc faults usually occur at the connection port of the charging device, the battery management system or the poor contact of the line. If such faults are not discovered in time, they may cause serious accidents such as fire. To this end, this paper discusses an early sign recognition and intelligent warning technology for arc faults in electric bicycle charging facilities. First, broadband high-frequency current and voltage sensors are used to capture the arc signals that may be generated during the charging process in real time. Then, wavelet transform is used to denoise and extract features of the signal to accurately separate the arc fault signal from the normal charging signal. Then, based on the long short-term memory neural network (LSTM) model, intelligent recognition and prediction of early signs are achieved. Finally, combined with edge computing technology, efficient fault warning is achieved. Experimental results show that the proposed system has an accuracy of about 95.6% in arc fault recognition. In terms of response time, the system can respond quickly within 0.2 seconds after the fault occurs and issue a warning in time. In terms of warning lead time, the system can provide an effective warning 0.3 seconds before the fault occurs to ensure the safety of charging facilities. In the above data conclusion, the experimental results verified the adaptability of the system under different battery types and noise environments, and demonstrated the great potential of this technology in the safety monitoring of electric bicycle charging facilities.

1. Introduction

At present, there are over 400 million electric bicycles in China, and with an annual growth rate of 10%, more than 200 million electric bicycles are being charged every day. In addition to the possibility of thermal runaway in lithium batteries themselves, DC arc faults caused by short circuits in electric bicycle charging lines, short circuits in connectors, cable insulation damage, poor contact, and other reasons are also the main culprits of electrical fires. More importantly, the fault arc itself during the charging process of electric bicycles is an important cause of thermal runaway in lithium batteries. At present, the existing charging facilities generally lack effective monitoring and warning mechanisms for arc faults. Therefore, how to accurately detect, cut off and warn in a timely manner before arc faults occur has become a key issue in improving the safety of electric bicycle charging. This study proposes an electric bicycle charging facility fault arc early warning technology based on high-frequency signal analysis and deep learning, aiming to improve charging safety and prevent potential safety hazards.

The main contribution of this paper is to propose a method for identifying arc faults in electric bicycle charging facilities that combines high-frequency signal analysis with deep learning. By capturing the current and voltage signals generated during the charging process in real time, using wavelet transform to denoise, extracting the characteristics of arc faults, and combining LSTM to perform dynamic fault identification and prediction. In addition, this paper also experimentally evaluates the adaptability of the system under different charging equipment and environmental conditions, proving that the system can maintain a high fault identification accuracy and a short response time in a variety of charging scenarios, effectively improving the safety of charging facilities.

This paper first introduces the safety hazards faced by electric bicycle charging facilities and the hazards of fault arcs, and explains the research background and significance. In the method stage, the design principles and methods of early sign identification and intelligent early warning technology of fault arcs are introduced. In the experimental part, the effectiveness of the proposed method is verified through multiple groups of experiments, covering the evaluation of fault arc identification accuracy, response time and early warning lead time. Finally, the results of this paper are discussed and future research directions are proposed.

2. Related Work

In the field of electrical equipment safety, arc monitoring and identification has become an important research direction. For example, the study by Tian et al. aimed to investigate the latest progress in arc fault detection, focusing on the role of artificial intelligence in this process. The purpose is to illustrate the advantages of artificial intelligence while identifying the current limitations [1]. AC series arcs were dangerous and may cause serious electrical fires and property losses. To address this problem, Wang et al. proposed a CNN-based arc detection model. The database used in the model collects data from eight different types of loads according to the IEC62606 standard [2]. Jiang et al. proposed a series arc fault detection and location algorithm for multi-load circuit topologies, specifically for branch arc faults and nonlinear power loads, to address the arc detection challenges caused by different load types and connections [3]. Due to the wide variety of arc faults caused by different load types, residential series arc fault detection becomes complex and challenging. Dang et al. selected and analyzed multiple typical loads, especially nonlinear and complex loads, for arc fault detection [4]. Shaloo et al. comprehensively reviewed various NDT techniques for wire and arc additive manufacturing and fusion welding, including laser ultrasound, acoustic emission using an airborne optical microphone, thermal imaging, and defect detection using oscilloscope monitoring of the welding current [5]. The arc detection algorithm proposed by Dołęgowski and Szmajda provides faster computation speed than existing algorithms and is cheaper to implement in arc detection equipment [6]. Fang et al. analyzed the current and voltage in the wire arc additive manufacturing process to detect possible anomalies and proposed a clustering-based anomaly detection method, which classified abnormal and normal instances into two independent clusters [7]. Although the methods proposed by these scholars have improved the accuracy of arc identification to a certain extent, they still face problems such as large data volume, poor real-time performance, and false alarms.

In response to the above-mentioned shortcomings, some researchers have begun to try to combine machine learning and big data technology to improve the accuracy and real-time performance of arc recognition. For example, in response to the difficulty in identifying subtle AC arcs in aviation cables, Liu et al. proposed an arc fault detection method that combines three-dimensional features and LSTM to perform arc fault analysis [8]. In order to solve the problem that the arc current of series arc faults in electric vehicle charging systems is difficult to accurately detect, Pan et al. proposed a series arc fault recognition method based on machine learning [9]. In order to study charging arc faults, Du established an arc fault model for electric vehicle charging systems, and obtained the change rules of the power battery pack terminal voltage, arc voltage and loop current under different battery charge states through computer simulation, providing a theoretical basis for the identification of electric vehicle charging loop arc faults [10]. Although these methods have achieved good results in specific environments, they may be affected by data noise and environmental interference in complex charging scenarios, resulting in a decrease in recognition accuracy. In order to solve this problem, this paper proposes an arc early sign recognition technology based on wavelet transform and LSTM network, which effectively avoids the false alarm and missed alarm problems in traditional methods and improves the performance of fault detection.

3. Method

3.1. Real-Time Acquisition of High-Frequency Signals

3.1.1. Acquisition Equipment and Configuration

The high-frequency current and voltage sensors used in this experiment have broadband frequency response capabilities and can be sampled in a frequency range of up to 1 MHz. The sampling frequency of the signal acquisition system is set to 1 kHz. This sampling frequency can ensure that the high-frequency signal fluctuations generated by the arc fault are captured and provide accurate raw data for subsequent signal analysis. The sampling theorem is shown in formula (1):

$$f_s \geq 2f_{max} \tag{1}$$

Among them, f_s represents the sampling frequency and f_{max} represents the maximum frequency of the signal. The current and voltage signals are collected in real time through sensors connected to the current loop and voltage loop of the electric bicycle charging facility, respectively. After normalizing the current signal with the sign function, determine whether the amplitudes of adjacent sampling points are the same. If they are the same, take 0, otherwise take 1. After calculation, take the sum and record it as the number of differences between adjacent points. Calculate the number of differences between adjacent points every two signal periods as a unit.The collected data is transmitted to the data acquisition card for subsequent processing and analysis. The design of the signal acquisition system can monitor the changes in current and voltage in real time without affecting the charging process, ensuring the system's rapid response to faults[11].

3.1.2. Signal Enhancement Algorithm Design

By improving the empirical wavelet transform and the improved singular value decomposition method, the current signal is enhanced to suppress the noise component in the current signal and effectively highlight the characteristics of the arc fault.

Improved empirical wavelet transform

The empirical wavelet transform (EWT) effectively separates the signal's load fundamental frequency component, wide-area fault arc component, non-stationary component and high-order harmonic component. The specific method is as follows:

1) The collected electrical signal is subjected to Fourier transform to obtain the Fourier spectrum of the signal, and it is normalized to obtain the Fourier spectrum in the interval [0,2π], and the support interval of the spectrum is defined in [0,π].

2) Assuming that the original signal consists of N components, select 2N maximum values in the spectrum, and use the minimum value between adjacent maximum values as the boundary frequency of signal division.

3) Calculate the average energy of each frequency band separately: when the average energy of each frequency band is greater than the average energy of the total frequency band, retain the frequency band.

4) Design a wavelet filter bank, after filtering, the signal can be separated into N groups of modal components.

3.1.3. Signal acquisition and storage

The signal acquisition process is ongoing, and the system continuously obtains current and voltage data and stores them in the database. In order to adapt to different charging environments, the system uses sliding window technology to slice the data collected each time. The data in each time window can be analyzed independently to ensure that the instantaneous changes of arc faults can be captured in real time.

The collected data mainly includes two aspects: current fluctuation and voltage fluctuation. These signal data usually include normal fluctuations in the charging process and abnormal fluctuations caused by arc faults. The system in this paper will transfer the raw data to the data processing module for denoising and feature extraction. In order to facilitate subsequent analysis and processing, all collected current and voltage signal data are stored in a database in a standardized format. The following Table 1 is a brief example of the data storage format:

Table 1. Data storage display

Timestamp (s)	Current (A)	Voltage (V)	Signal Type
0.001	2.01	48	Normal Charging
0.002	2.012	48.002	Normal Charging
0.003	2.05	48.05	Normal Charging
0.004	2.1	48.1	Normal Charging
0.005	3	50	Fault Occurred
0.006	4.5	51	Fault Occurred
0.007	4.8	52	Fault Occurred
0.008	3.2	50.2	Fault Recovery

The table shows the data format of the signal acquisition, including timestamp, current, voltage, and signal type (such as normal charging or fault occurrence). When a fault occurs, the current and voltage increase rapidly, and these fluctuations are significantly different from the stable fluctuations during normal charging. This data storage method can efficiently manage a large amount of real-time data and provide sufficient raw data support for subsequent fault detection and early warning analysis.

3.2. Arc Signal Feature Extraction

3.2.1. Data Preparation and Spatiotemporal Feature Construction

In order to extract effective features from high-frequency signals, the original signal must first be denoised. When an arc fault occurs, current and voltage signals are usually accompanied by noise. Therefore, denoising is the first step to ensure signal clarity. This paper uses wavelet transform for signal denoising because wavelet transform has good time-frequency localization characteristics and can effectively extract features from signals at multiple scales. The formula is as follows (2):

$$W_{\psi}(a,b) = \frac{1}{\sqrt{|a|}} \int_{-\infty}^{\infty} x(t)\psi\left(\frac{t-b}{a}\right)dt \tag{2}$$

Among them, $x(t)$ represents the original signal, ψ is the mother wavelet. Wavelet transform decomposes the signal into approximate coefficients and detail coefficients through multi-layer decomposition. The approximate coefficients represent the low-frequency part of the signal, while the detail coefficients capture the high-frequency fluctuations of the signal. In arc fault signals, the detail coefficients contain the key features of the fault, especially in the high-frequency fluctuation part. Therefore, we use wavelet transform denoising to remove the noise part in the original signal and retain the high-frequency components related to the fault in the signal [12].

3.2.2. Time Domain and Frequency Domain Feature Extraction

In the denoised signal, arc faults are usually manifested as rapid changes in current and voltage. To better understand the impact of arc faults on signals, Figure 1 shows the changes in current and voltage for normal charging (no fault) and arc fault signals.

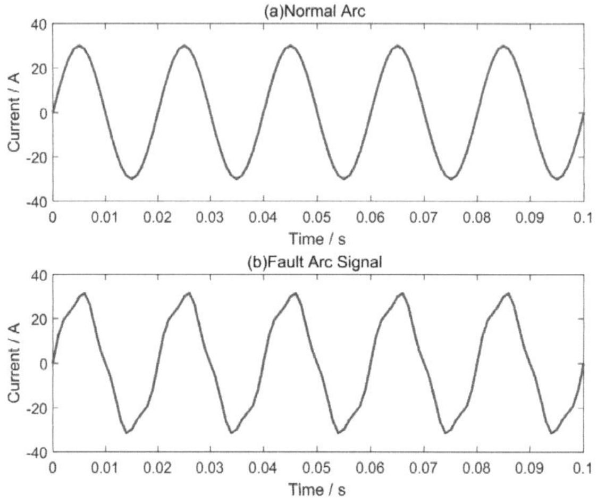

Figure 1. Arc waveform

Figure 1(a) shows the steady fluctuation of the current signal during normal charging, while Figure 1(b) shows the fluctuation of the current signal after an arc fault occurs. The instantaneous increase and fluctuation caused by the fault are clearly visible. In order to extract effective features, this paper performs feature analysis in both the time domain and the frequency domain.

Time domain feature extraction: In the time domain, arc faults are usually manifested as sudden changes or abnormal fluctuations in current and voltage. Common time domain features include peak value, mean value, standard deviation, volatility, etc. These features can reflect the amplitude change of the signal, especially when a fault occurs, the fluctuation of the signal shows obvious abnormality. By calculating the time domain features, the obvious changes of arc faults can be effectively captured.

Frequency domain feature extraction: In the frequency domain, arc faults produce obvious high-frequency noise. In order to extract these high-frequency features, this paper uses fast Fourier transform to convert time domain signals into frequency domain signals. By analyzing the high-frequency components in the spectrum, the characteristic frequency of the arc fault can be identified. In addition, the signal's spectral energy, spectral entropy and other features can also be calculated. These features usually show large energy changes during arc faults and can provide more clues for fault detection.

3.3. Time Series Data Learning and Fault Identification

3.3.1. Application of LSTM

In order to effectively process time series signals, this paper adopts the LSTM model for arc fault identification. The input of the LSTM model is the current and voltage time series data that has been processed by denoising and feature extraction. By learning these data, LSTM can extract the dynamic change rules of arc faults. For newly input time series data, the model judges and predicts fault signals based on the learned rules. The performance of LSTM in dynamic signal identification can accurately capture the precursor changes of arc faults and significantly improve the accuracy of fault identification. Its loss function is shown in formula (3):

$$L = \frac{1}{N}\sum_{i=1}^{N}(y_i - \hat{y}_i)^2 \quad (3)$$

Among them, y_i represents the actual label, $\hat{y}_i$ is the model's prediction, and N represents the total number. Through this formula, the LSTM model can gradually learn the dynamic change law of arc faults[13].

3.3.2. Model Training and Validation

The training process of the LSTM model includes three main steps: data preprocessing, model construction, and optimization. In the data preprocessing stage, the collected current and voltage signals are first normalized to ensure the uniformity and stability of the data. Then, the time series data is divided into multiple time periods using the split window method. The data in each time period contains the fluctuation information of current and voltage and is used as the input sample.

In the model building stage, the architecture design of the LSTM network includes an input layer, multiple LSTM layers, and an output layer. Its architecture is shown in Figure 2:

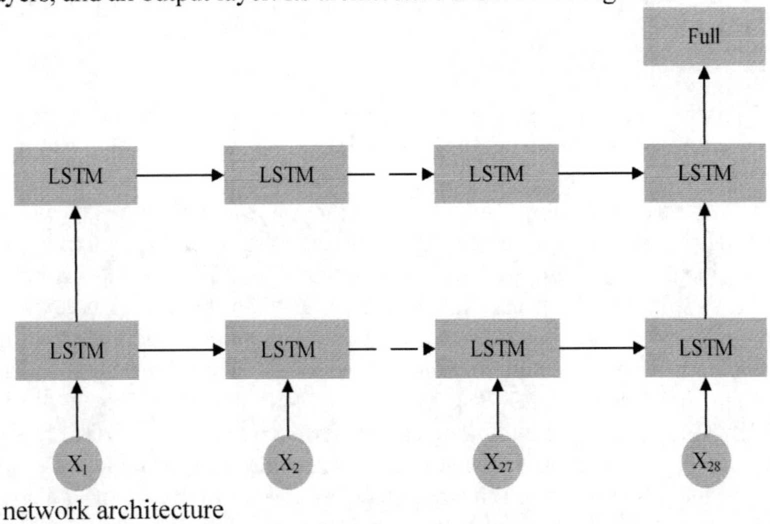

Figure 2. LSTM network architecture

The input layer is responsible for receiving the time series data of current and voltage. The LSTM layer processes the time series data through multiple neural units and generates prediction results at the output layer. In order to improve the accuracy of the model, this paper also introduces the Dropout technology to prevent the model from overfitting during the training process [14].

3.4. Edge Computing Realizes Real-Time Warning

3.4.1. Edge Computing Design

In this study, we proposed a real-time fault warning system based on edge computing. The system consists of data collection layer, edge computing layer and cloud service layer. The data acquisition layer collects real-time operating data such as current and voltage through high-frequency current and voltage sensors installed on the charging facilities. These data are first processed and analyzed by edge computing devices to identify fault signals and issue warnings. If there is a possibility of an arc fault, the system immediately generates a warning on the edge device and sends the results to the cloud service layer through the network for further data storage and processing.

The edge computing layer is responsible for local data processing, fault identification and early warning. The core task of this layer is to analyze the collected data in real time, determine whether there is an arc fault risk, and respond quickly. The cloud service layer is mainly responsible for long-term data storage, model training and system management, and provides subsequent optimization and upgrades of the system.

3.4.2. Implementation of Edge Computing

Traditional fault detection systems usually need to transmit all data to the cloud for processing, which causes transmission delays, especially when the network is unstable or the bandwidth is limited. Edge computing significantly reduces transmission delays by processing data near the data source. In this system, the edge computing device is equipped with an efficient microprocessor and small hardware adapted to deep learning, which can quickly process current and voltage signals.

In terms of signal processing, the edge computing device first denoises the signal and then extracts key features. The extracted data is input into the trained LSTM model for fault identification. The LSTM model determines the risk of arc faults by analyzing time series data, and generates an early warning when a fault is detected, sending information in a timely manner through the local network to ensure that the problem can be handled as soon as possible.

4. Results and Discussion

4.1. Experimental Preparation

Hardware configuration: The fault arc signal monitored by this project is an analog signal with a high signal frequency. Special analog-to-digital conversion devices are often used to design signal acquisition circuits. This prototype uses STMicroelectronics' general-purpose control chip STM32F103C8T6 as a sampling chip. A dual-channel monitoring solution is designed for the charging port. The main channel captures the current ripple characteristics, and the auxiliary channel tracks the transient voltage fluctuations. Equipped with Advantech UNO-2484G edge computing gateway, its GD32F470VGT6 microprocessor from GigaDevice can achieve feature value extraction within 12ms and provide preprocessing data for the LSTM prediction module. The display module is a sub-component of the human-computer interaction part, which is used to display the series information of the prototype. The 0.91-inch OLED display screen used in the prototype of this project has a resolution of 128*64, and the data communication interface is an IIC interface. As part of the power box protection device, the monitoring results need to be uploaded to the cloud platform through the user data transmission system through the electrical fire monitoring system to achieve remote macro monitoring. The communication module of this prototype uses the RS485 communication interface.

Software development: A layered architecture is built based on Python 3.9. The data acquisition layer used a multi-threaded buffer mechanism to ensure sampling continuity. The signal processing layer integrated the wavelet packet decomposition algorithm of PyWavelets 1.4.1 (using the db8 basis function) and the Savitzky-Golay filter of SciPy 1.9.3. Model training is implemented using CudnnLSTM of TensorFlow 2.10, and the training efficiency is optimized through dynamic learning rate adjustment (initial lr=0.001, attenuation coefficient 0.1/5epoch).

Test samples: Typical charging equipment of the top five electric bicycle brands in terms of market share is selected, covering three categories: ternary lithium battery, lithium iron phosphate and Chaowei lead-acid battery. Two extreme working conditions of 40°C/90%RH high temperature and high humidity and -20°C low temperature are specially set to simulate the actual user environment.

Fault simulation: Several common arc fault types are simulated, including poor contact faults, short circuit faults and faults caused by battery aging, in order to examine the system's ability to identify different arc faults.

4.2. Experimental Analysis

4.2.1. Fault Arc Identification Accuracy Evaluation Experiment

This experiment simulates the arc fault that may occur during the charging process of electric bicycles and studies the early fault sign recognition technology based on high-frequency signal analysis and deep learning. First, the current and voltage signals are denoised using wavelet transform, and then the LSTM model is used to identify and predict the fault signals. The experiment uses different signal environments, including normal charging signals and noisy arc fault signals, to evaluate the system's performance in recognition accuracy and real-time performance. By comparing with the traditional threshold detection method, the effectiveness and advantages of the system are further verified. The specific data is shown in Figure 3:

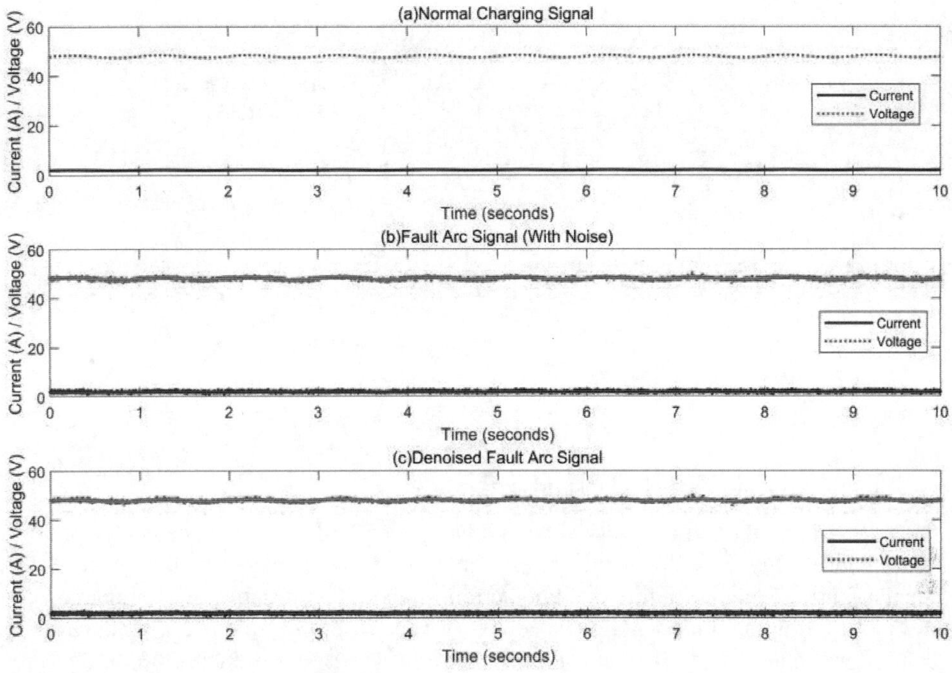

Figure 3. Arc fault identification accuracy evaluation

Figure 3 (a-c) shows the normal charging signal, the fault arc signal (with noise) and the denoised fault arc signal. In the simulated arc fault environment, the arc recognition accuracy of the LSTM model is about 95.6%, which is significantly higher than the traditional monitoring method based on current and voltage threshold methods (about 85.3%). The wavelet transform effectively removes the noise in the arc signal and retains the main features, while the LSTM model accurately captures the dynamic evolution characteristics. The system shows strong adaptability under a variety of charging equipment and environmental conditions, providing a reliable solution for the safe monitoring of electric bicycle charging facilities.

4.2.2. Response Time and Warning Lead Time Evaluation Experiment

In order to evaluate the response time and warning ability of electric bicycle charging facilities when a fault arc occurs, two experiments are designed. Figure 4 simulates a sudden fault, that is, an arc fault occurs during the charging process, and the current and voltage fluctuate rapidly. Figure 5 simulates a gradual fault, that is, the current and voltage increase slowly and a warning signal is issued in advance. Through these two experiments, the response time of different types of faults and the warning lead time of the system are evaluated:

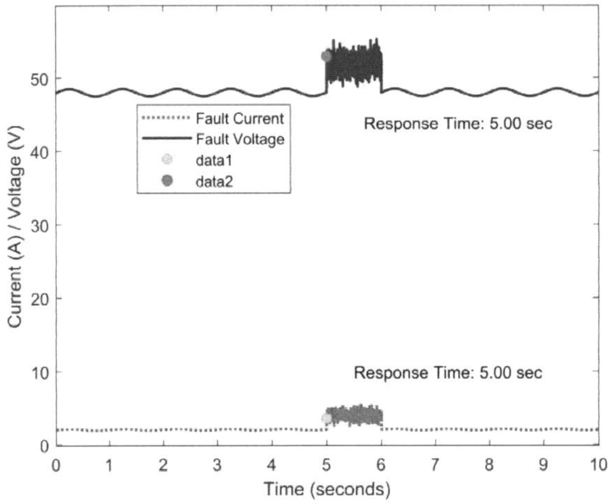

Figure 4. Sudden fault signal and its response time

In this experiment, the fault arc signal occurs between 5 and 6 seconds, causing a sharp increase in current and voltage. Figure 4 shows the changes in current (red curve) and voltage (blue curve) before and after the fault. Through the threshold detection method, the system issues an early warning when the current exceeds 3A and the voltage exceeds 50V. The response time is 5.2 seconds for current and 5.1 seconds for voltage. The green and purple marks in the figure indicate the time points when the current and voltage reaches the threshold.

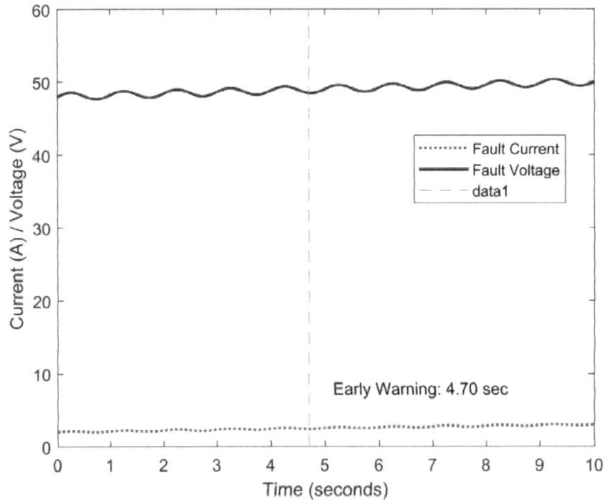

Figure 5. Progressive fault signal and its warning lead time

Figure 5 simulates a progressive arc fault, where the current and voltage begin to increase slowly at 5 seconds. Through threshold detection, the system predicts the fault occurrence time and issues a warning 0.3 seconds in advance. Figure 5 shows the gradual increase trend of current (red curve) and voltage (blue curve). The green dotted line indicates the time point when the system issues a warning, ensuring that the system can provide timely warnings before the fault occurs.

Through these two experiments, the system based on wavelet transform and LSTM network showed high accuracy and response speed in the early identification and warning of fault arc. In the sudden fault experiment, the system is able to respond accurately and issue a warning within 0.2 seconds; in the gradual fault experiment, the system is able to issue a warning 0.3 seconds in advance, ensuring the safety of the charging facilities. This shows that the warning technology based on signal

Journal of Physics: Conference Series **3036** (2025) 012009 doi:10.1088/1742-6596/3036/1/012009

analysis and deep learning can significantly improve the safety performance of electric bicycle charging facilities, effectively identify fault risks in advance and avoid potential accidents.

4.2.3. Adaptive Adjustment Mechanism Evaluation Experiment

This experiment aims to evaluate the adaptability of the arc fault identification system of electric bicycle charging facilities under different charging equipment and environmental conditions. By simulating five scenarios, changing the battery type (lithium battery, lead-acid battery, supercapacitor) and electrical noise environment (no noise, low noise, high noise), the system's fault identification accuracy, response time and warning lead time under these different conditions are tested. The experiment analyzes the performance and adaptability of the system in actual applications by comparing the experimental data under different scenarios. The specific data can be seen in Table 2:

Table 2. Evaluation of adaptive adjustment mechanisms

Scenario	Battery Type	Noise Environment	Fault Detection Accuracy (%)	Response Time (s)	Early Warning Time (s)
1	Lithium Battery	No Noise	96.5	0.22	0.3
2	Lead-Acid Battery	Low Noise	94.2	0.23	0.28
3	Lithium Battery	High Noise	91.5	0.25	0.32
4	Super Capacitor	No Noise	97	0.2	0.35
5	Lead-Acid Battery	High Noise	89.8	0.27	0.29

The experimental results show that the system has high stability and adaptability under different battery types. In a noise-free environment, the fault identification accuracy of lithium batteries and supercapacitors is 96.5% and 97.0%, respectively, while in a high-noise environment, the accuracy drops slightly to 91.5%. The response time of the system is maintained at around 0.22 seconds in both noise-free and low-noise environments, and 0.25 seconds in a high-noise environment. The warning lead time is maintained between 0.30 and 0.35 seconds, ensuring timely and effective fault warning. Overall, the system based on wavelet transform and LSTM network performs stably in different environments, providing reliable fault detection and warning capabilities for electric bicycle charging facilities.

4.3. Experimental Discussion

In the arc fault identification accuracy evaluation experiment, the system based on wavelet transform and LSTM network showed high recognition ability and could effectively detect arc faults. The response time and warning lead time evaluation experiment showed that the system responded quickly and could issue warnings in time before the fault occurred, thus ensuring safety. In the adaptive adjustment mechanism evaluation experiment, the system is able to maintain good performance under different battery types and noise environments. Although the high noise environment would have a certain impact on the system performance, the overall performance is still stable. These results show that the system has strong adaptability and reliability and can meet a variety of practical application needs.

5. Conclusion

This paper studies the early sign recognition and intelligent warning technology of arc faults in electric bicycle charging facilities, and proposes a method combining wavelet transform and LSTM network. Through real-time acquisition, denoising and feature extraction of high-frequency current and voltage signals, the system can accurately identify arc faults and issue early warnings, significantly improving the safety of charging facilities. The experimental results show that the fault recognition accuracy of the system reaches about 95.6%, and it can respond quickly within 0.2 seconds after the fault occurs and issue a warning 0.3 seconds in advance, showing strong real-time and adaptability. Despite this, there are still some limitations in this study, mainly reflected in the simulation of experimental data and the need for further optimization of deep learning models. Future research can verify the feasibility of the technology in the actual charging environment, explore the impact of different fault types and environmental factors on system performance, and combine big data analysis with cloud platform monitoring technology to further improve the intelligence level and universality of the system and promote the widespread application of electric bicycle charging safety monitoring.

Reference

[1] Tian C., Xu Z., Wang L., et al. Arc fault detection using artificial intelligence: Challenges and benefits [J]. Mathematical Biosciences and Engineering, 2023, 20(7): 12404-12432.

[2] Wang Y., Hou L., Paul K. C., et al. ArcNet: Series AC arc fault detection based on raw current and convolutional neural network [J]. IEEE Transactions on Industrial Informatics, 2021, 18(1): 77-86.

[3] Jiang J., Li W., Wen Z., et al. Series arc fault detection based on random forest and deep neural network [J]. IEEE Sensors Journal, 2021, 21(15): 17171-17179.

[4] Dang H. L., Kim J., Kwak S., et al. Series DC arc fault detection using machine learning algorithms [J]. IEEE Access, 2021, 9(1): 133346-133364.

[5] Shaloo M., Schnall M., Klein T., et al. A review of non-destructive testing (NDT) techniques for defect detection: application to fusion welding and future wire arc additive manufacturing processes [J]. Materials, 2022, 15(10): 3697-3714.

[6] Dołęgowski M., Szmajda M. A novel algorithm for fast DC electric arc detection [J]. Energies, 2021, 14(2): 288-296.

[7] Fang J., Wang Z., Liu W., et al. A new particle swarm optimization algorithm for outlier detection: Industrial data clustering in wire arc additive manufacturing [J]. IEEE Transactions on Automation Science and Engineering, 2022, 21(2): 1244-1257.

[8] Liu X., Huang D., Jing T., et al. Detection of AC arc faults of aviation cables based on HIW three-dimensional features and CNN-LSTM neural network [J]. IEEE Access, 2022, 10(3): 106958-106971.

[9] Pan G.X., Pei L.W., Li X.Y., et al. Intelligent identification method of series arc fault in electric vehicle charging system [J]. Journal of Electric Power System and Automation, 2023, 35(10):107-114.

[10] Du X.Y. Modeling and simulation of arc fault in electric vehicle charging system [J]. Automobile Practical Technology, 2023, 48(10):1-6.

[11] Zhao S., You G., Hou X., et al. A Spatial Location Method for DC Series Arc Faults Based on RSSI and Bayesian Regularization Neural Network [J]. IEEE Sensors Journal, 2021, 21(24): 27868-27877.

[12] Xia C., Pan Z., Li Y., et al. Vision-based melt pool monitoring for wire-arc additive manufacturing using deep learning method [J]. The International Journal of Advanced Manufacturing Technology, 2022, 120(1): 551-562.

[13] Alcaraz J. Y., Sharma A., Tjahjowidodo T. Predicting porosity in wire arc additive manufacturing (WAAM) using wavelet scattering networks and sparse principal component analysis [J]. Welding in the World, 2024, 68(4): 843-853.

[14] Tang J., Gao W. A novel cascaded H-bridge photovoltaic inverter with flexible arc suppression function [J]. Global Energy Interconnection, 2024, 7(4): 513-527.

AUTHOR INDEX